I0465724

PUBLISHER COMMENTARY

We print NASA's handbooks and standards for the convenience of those that use them on a daily basis. We print all of these a full 8 ½ by 11 with large text so they are easy to read. Yes, color books are expensive to print so unless the information relies on the use of color for proper interpretation or understanding, we print most books in black and white to keep the cost down. All these documents are available for download for free from NASA, however printing them all over a network printer would take days.

Why buy a book you can download free? We print this so you don't have to.

All these books are available for free download from the government web site. Some are available only in electronic media. Some online docs are missing pages or barely legible.

We at 4th Watch Publishing are former government employees, so we know how government employees actually use the standards. When a new standard is released, an engineer prints it out, punches holes and puts it in a 3-ring binder. While this is not a big deal for a 5 or 10-page document, many NIST documents are over 100 pages and printing a large document is a time-consuming effort. So, an engineer that's paid $75 an hour is spending hours simply printing out the tools needed to do the job. That's time that could be better spent doing engineering. We publish these documents so engineers can focus on what they were hired to do – engineering. It's much more cost-effective to just order the latest version from Amazon.com

If there is a standard you would like published, let us know. Our web site is www.usgovpub.com

www.usgovpub.com

List of Other NASA Publications Available on Amazon.com:

NASA-STD-5001B	Structural Design and Test Factors of Safety for Spaceflight Hardware
NASA-STD-5006A	General Welding Requirements for Aerospace Materials
NASA-STD-5008B	Protective Coating of Carbon Steel, Stainless Steel, and Aluminum on Launch Structures, Facilities, and Ground Support Equipment
NASA-STD-5009A	Nondestructive Evaluation Requirements for Fracture-Critical Metallic Components
NASA-STD-5012B	Strength and Life Assessment Requirements for Liquid-Fueled Space Propulsion System Engines
NASA-STD-5019A	Fracture Control Requirements for Spaceflight Hardware
NASA-STD-5005D	Standard for The Design and Fabrication of Ground Support Equipment
NASA-HDBK-8739.21	Workmanship Manual for Electrostatic Discharge Control
NASA-HDBK 8739.23A	NASA Complex Electronics Handbook for Assurance Professionals (Color)
NASA-HDBK-8719.14	Handbook for Limiting Orbital Debris (Color)
NASA-HDBK-8709.22	Safety and Mission Assurance Acronyms, Abbreviations, and Definitions
NASA-HDBK-7009	NASA Handbook for Models and Simulations: An Implementation Guide For NASA-STD-7009 (Color)
NASA-HDBK-8739.19-2	Measuring and Test Equipment Specifications NASA Measurement Quality Assurance Handbook – Annex 2
NASA-HDBK-8739.19-3	Measurement Uncertainty Analysis Principles and Methods NASA Measurement Quality Assurance Handbook – Annex 3
NASA-HDBK-8739.19-4	Estimation and Evaluation of Measurement Decision Risk NASA Measurement Quality Assurance Handbook – Annex 4
NASA RCM	Reliability-Centered Maintenance Guide for Facilities and Collateral Equipment

NASA HANDBOOK

Measurement Uncertainty Analysis Principles and Methods

NASA Measurement Quality Assurance Handbook – ANNEX 3

Measurement System Identification: **Metric**

July 2010

National Aeronautics and Space Administration
Washington DC 20546

This page intentionally left blank.

DOCUMENT HISTORY LOG

Status	Document Revision	Approval Date	Description
Baseline		2010-07-13	Initial Release *(JWL4)*
	Revalidated	2018-03-01	Baseline revalidated.

This document is subject to reviews per Office of Management and Budget Circular A-119, Federal Participation in the Development and Use of Voluntary Standards (02/10/1998) and NPD 8070.6, Technical Standards (Paragraph 1.k).

FOREWORD

This publication provides principles and methods for the analysis and management of the uncertainty that exists in measurements. The principles and methods described herein support Agency objectives for assuring measurement accuracy and are applicable for use in all NASA functions for which measurements and decisions based on measurements are involved.

This Annex to NASA-HDBK 8739.19 is approved for use by NASA Headquarters and NASA Centers, including Component Facilities. This document may be referenced on contracts as a guidance or training publication.

Comments and questions concerning the contents of this publication should be referred to the National Aeronautics and Space Administration, Director, Safety and Assurance Requirements Division, Office of Safety and Mission Assurance, Washington, DC 20546.

Requests for information, corrections, or additions to this NASA HDBK shall be submitted via "Feedback" in the NASA Technical Standards System at http://standards.nasa.gov or to National Aeronautics and Space Administration, Director, Safety and Assurance Requirements Division, Office of Safety and Mission Assurance, Washington, DC 20546.

Bryan O'Connor
Chief, Safety and Mission Assurance

13 JULY 2010
Approval Date

This page intentionally left blank.

TABLE OF CONTENTS

CHAPTER 6: MULTIVARIATE MEASUREMENTS   59

CHAPTER 7: MEASUREMENT SYSTEMS   79

LIST OF FIGURES

LIST OF TABLES

ACRONYMS AND ABBREVIATIONS

ADC	Analog to Digital Converter
AOP	Average-over-Period
BOP	Beginning-of-Period
BSL	Best-fit Straight Line
CMR	Common-Mode Rejection
CMRR	Common-Mode Rejection Ratio
CMV	Common-Mode Voltage
CTE	Coefficient of Thermal Expansion
DAC	Digital to Analog Converter
DMM	Digital Multimeter
EOP	End-of-Period
FS	Full Scale
FSI	Full Scale Input
FSO	Full Scale Output
FSR	Full Scale Range
LSB	Least Significant Bit
LSD	Least Significant Digit
GUM	Guide to the Expression of Uncertainty in Measurement
ISO	International Organization for Standardization
MTE	Measuring and Test Equipment
NMRR	Normal-Mode Rejection Ratio
NMV	Normal-Mode Voltage
OOT	Out of Tolerance
PDF	Probability Density Function
ppb	Parts per billion
ppm	Parts per million
RDG	Reading
RF	Radio Frequency
RH	Relative Humidity
RO	Rated Output
RSS	(1) Root-sum-square method of combining values. (2) Residual sum of squares in regression analysis.
RTI	Referred to Input
RTO	Referred to Output
SU	Subject Unit
TRS	Transverse Rupture Strength
UUT	Unit Under Test
VAC	Alternating Current Volts
VDC	Volts Direct Current

PREFACE

Measurement uncertainty is the doubt that exists about a measurement's result. Every measurement - even the most careful - *always* has a margin of doubt. Evaluating the uncertainty in the measurement process determines the "goodness" of a measurement.

This Handbook provides tools for estimating the quality of measurements. Measurement uncertainty is an estimation of the potential error in a measurement result that is caused by variability in the equipment, the processes, the environment, and other sources. Every element within a measurement process contributes errors to the measurement result, including characteristics of the item being tested. Evaluation of the measurement uncertainty characterizes what is reasonable to believe about a measurement result based on knowledge of the measurement process. It is through this process that credible data can be provided to those responsible for making decisions based on the measurements.

In this context, it becomes apparent the more critical the application, the greater the need for measurement quality assurance. Measurement uncertainty analysis can be used to mitigate risks associated with noncompliance of specifications and/or requirements which are validated through measurement. Although the tools are available, often the overall uncertainty encountered during the measurement process is not assessed, controlled, or even fully understood. The principles and methods recommended in this Handbook may be used as the fundamental building blocks for a quality measurement program. From this foundation, good measurement data can support better decisions.

Ensuring reliable and accurate products and services justifies a measurement assurance program as a cost benefit - providing the assurance of safety through measurement quality makes it imperative.

A lack of standardization for quantified measurement uncertainty estimation often causes disagreements and confusion in trade, scientific findings, and legal issues. The principles and methods contained in this Handbook are based, and in some instances, expand on the International Organization for Standardization (ISO) *Guide to the Expression of Uncertainty in Measurement* (GUM), the international standardized approach to estimating uncertainty. ANSI/NCSL Z540.2-1997 (R2007), *U.S. Guide to the Expression of Uncertainty in Measurement* (U.S. Guide), is the U.S. adoption of the ISO GUM. Additional guidance on estimating measurement uncertainty is available in many engineering discipline-specific voluntary consensus standards and complimentary documents. However, for consistent results, it is imperative that the quantification of measurement uncertainty be based on the ISO GUM.

ACKNOWLEDGEMENTS

The principal authors of this publication are Ms. Suzanne Castrup and Dr. Howard T. Castrup of Integrated Sciences Group (ISG), Inc. of Bakersfield, CA under NASA Contract.

Additional contributions and critical review were provided by:

Mihaela Fulop,
SGT, NASA Glenn Research Center

Dr. Dennis Jackson
U.S. Naval Warfare Assessment Station

Scott M. Mimbs
NASA Kennedy Space Center

Dr. Dennis Dubro
Pacific Gas and Electric Company, CA

Dr. Li Pi Su
Army Primary Standards Lab, AL

Mark Kuster
BWXT Pantex, TX

Paul Reese
Bionetics, Inc., NASA Kennedy Space Center, FL

William Hinton
Nextera Energy, NH

James Wachter
MEI Co., NASA Kennedy Space Center, FL

This Publication would not exist without the recognition of a priority need by the NASA Metrology and Calibration Working Group and funding support by NASA/HQ/OSMA and NASA/Kennedy Space Center.

EXECUTIVE SUMMARY

Measurements are an important aspect of decision making, communicating technical information, establishing scientific facts, monitoring manufacturing processes and maintaining human and environmental health and safety. Consequently, industries and governments spend billions of dollars annually to acquire, install and maintain measurement and test equipment (MTE).

The more critical the application, the greater the need for measurement quality assurance. MTE accuracy is a key aspect of measurement quality. However, the overall uncertainty encountered during the measurement process is not often assessed and controlled.

The assessment and control of measurement uncertainty presupposes the ability to develop reliable uncertainty estimates. This document provides an in-depth coverage of key aspects of measurement uncertainty analysis and detailed procedures needed for developing such estimates.

Chapter 1 presents the purpose and scope of this document and discusses principal differences between "classical" engineering methods and more recent methods developed to provide an international consensus for the expression of uncertainty in measurement.

Chapters 2 and 3 provide foundational concepts and methods for estimating measurement uncertainty. Key concepts and methods are summarized below. Chapter 4 discusses how manufacturer specifications are obtained, interpreted and applied in uncertainty estimation.

Chapters 5 through 7 present procedures for implementing key concepts and methods, using detailed direct measurement, multivariate measurement and measurement system examples. Chapter 8 provides guidance and illustrative examples for estimating the uncertainty in the measurement result obtained from four common calibration scenarios. Chapter 9 presents an advanced topic for estimating uncertainty growth over time.

Appendix A provides definitions for terms employed throughout this document. The terms and definitions are designed to be understood across a broad technology base. Where appropriate, terms and definitions have been taken from internationally recognized standards and guidelines.

Appendices B through D provide in-depth development of concepts and methods described in Chapters 2 and 3. Appendix E provides an advance topic on applying Bayesian analysis to estimate unit-under-test (UUT) and MTE attribute biases and in-tolerance probabilities during calibration. Appendices F through I provide additional analysis examples.

Key Uncertainty Analysis Concepts and Methods

Measurement Error and Uncertainty

A measurement is a process whereby the value of a quantity is estimated. All measurements are accompanied by error. Our lack of knowledge about the sign and magnitude of measurement error is called measurement uncertainty. Measurement errors are random variables that follow probability distributions. A measurement uncertainty estimate is the characterization of what we know statistically about the measurement error. Therefore, a measurement result is only complete when accompanied by a statement of the uncertainty in that estimate.

Uncertainty Analysis

Uncertainty is calculated to support decisions based on measurements. Therefore, uncertainty estimates should realistically reflect the measurement process. In this regard, the person tasked with conducting an uncertainty analysis must be knowledgeable about the measurement process under investigation.

To facilitate this endeavor, the measurement process should be described in writing. Such documentation should clearly specify the measurement equipment used, the environmental conditions during measurement, and the procedure used to obtain the measurement.

The general uncertainty analysis procedure consists of the following steps:

1. Define the Measurement Process
2. Identify the Error Sources and Distributions
3. Estimate Uncertainties
4. Combine Uncertainties
5. Report the Analysis Results

The first step in any uncertainty analysis is to identify the physical quantity whose value is estimated via measurement. This quantity, sometimes referred to as the "measurand," may be a directly measured value or indirectly determined through the measurement of other variables. It is also important to describe the test setup, environmental conditions, technical information about the instruments, reference standards, or other equipment used and the procedure for obtaining the measurement(s). This measurement process information is used to identify potential sources of error.

Measurement process errors are the basic elements of uncertainty analysis. Once these fundamental error sources have been identified, then the appropriate distributions are selected to characterize the statistical nature of the measurement errors.

With a basic understanding of error distributions and their statistics, we can estimate uncertainties. The spread in an error distribution is quantified by the distribution's standard deviation, which is the square root of the distribution variance. Measurement uncertainty is equal to the standard deviation of the error distribution. There are two approaches to estimating measurement uncertainty. Type A estimates involve data sampling and analysis. Type B estimates use technical knowledge or recollected experience of measurement processes.

Because uncertainty is equal to the square root of the distribution variance, uncertainties from different error sources can be combined by applying the "variance addition rule." Variance addition provides a method for correctly combining uncertainties that accounts for correlations between error sources. When uncertainties are combined, it is also important to estimate the degrees of freedom for the combined uncertainty. Generally speaking, degrees of freedom signify the amount of information or knowledge that went into an uncertainty estimate.

Reporting Uncertainty

When reporting the results of an uncertainty analysis, the following information should be included:

1. The estimated value of the quantity of interest and its combined uncertainty and degrees of freedom.

2. The mathematical relationship between the quantity of interest and the measured components (applies to multivariate measurements).

3. The value of each measurement component and its combined uncertainty and degrees of freedom.

4. A list of the measurement process uncertainties and associated degrees of freedom for each component, along with a description of how they were estimated.

5. A list of applicable correlation coefficients, including any cross-correlations between component uncertainties.

It is also a good practice to provide a brief description of the measurement process, including the procedures and instrumentation used, and additional data, tables and plots that help clarify the analysis results.

This page intentionally left blank.

CHAPTER 1: INTRODUCTION

Concepts and methods presented in this document are consistent with those found in the International Organization for Standardization (ISO) *Guide to the Expression of Uncertainty in Measurement* (GUM).[1]

Uncertainty is calculated to support decisions based on measurements. Therefore, uncertainty estimates should realistically reflect the measurement process. In this regard, the person tasked with conducting an uncertainty analysis must be knowledgeable about the measurement process under investigation.

> **Note:** In this document, the terms standard uncertainty and uncertainty are used interchangeably.

To facilitate this endeavor, the measurement process should be described in writing. This write-up should clearly specify the measurement equipment used, the environmental conditions during measurement, and the procedure used to obtain the measurement.

1.1   Purpose

While the GUM provides general rules for analyzing and communicating measurement uncertainty, it does not provide detailed procedures or instructions for evaluating specific measurement processes.[2] In addition, new methods have been developed over the past several years that enhance the methodology of the GUM.

This document provides a recommended practice that clearly explains key concepts and principles for estimating and reporting measurement uncertainty. This document also includes advanced methods that extend the GUM's guidance on estimating measurement uncertainty.

1.2   Scope

The analysis methods outlined in this document provide a comprehensive approach to estimating measurement uncertainty. Basic guidelines are presented for estimating the uncertainty in the value of a quantity for the following measurement alternatives:

- Direct Measurements – The value of a quantity is obtained directly by measurement and not determined indirectly by computing its value from the values of other variables or quantities.

- Multivariate Measurements – The value of the quantity is based on measurements of more than one attribute or quantity.

- Measurement Systems – The value of a quantity is measured with a system comprised of component modules arranged in series.

The structured, step-by-step uncertainty analysis procedures described herein address the important aspects of identifying measurement process errors and using appropriate error models and error distributions. Advanced topics cover estimating degrees of freedom for Type B

---

[1] Throughout this document, the term GUM refers to ISO *Guide to the Expression of Uncertainty in Measurement* and ANSI/NCSL Z540-2-1997, the U.S. *Guide to the Expression of Uncertainty in Measurement*.

[2] See section 1.4 of the GUM.

uncertainties, uncertainty analysis for alternative calibration scenarios, uncertainty growth over time and Bayesian analysis.

Examples contained in the main body of this document provide detailed step-by-step analysis procedures that re-enforce important principles and methods. Analysis examples included in the appendices address real-world measurement scenarios and follow a standardized format to clearly convey the necessary information and concepts used in each analysis.

1.3   Background

The GUM was developed to provide an international consensus for the expression of uncertainty in measurements. This entailed the development of an unambiguous definition of measurement uncertainty and the application of rigorous mathematical methods for uncertainty estimation.

Over the past twenty years or so, various uncertainty analysis standards, guides and books have been published by engineering organizations. Examples of uncertainty analysis standards and other published material commonly used in the U.S. engineering community are listed below.

- *Test Uncertainty*, ASME PTC 19.1-1998 (reaffirmed 2004).

- *Measurement Uncertainty for Fluid Flow in Closed Conduits*, ANSI/ASME MFC-2M-1983 (reaffirmed 2001).

- *Assessment of Wind Tunnel Data Uncertainty*, AIAA Standard S-071-1995.

- Dieck, R.H.: *Measurement Uncertainty Methods and Applications*, 3rd Edition, ISA 2002.

- Coleman, H. W. and Steele, W. G.: *Experimentation and Uncertainty Analysis for Engineers*, 2nd Edition, John Wiley & Sons, 1999.

Although many of these uncertainty analysis references have been updated or reaffirmed in recent years, the methods they espouse are distinctly different from those presented in the GUM. Consequently, confusion persists in the reporting and comparison of uncertainty estimates across technical organizations and disciplines.

The methods and concepts presented in this document follow the GUM and are based on the properties of measurement error and the statistical nature of measurement uncertainty. Publications consistent with the GUM are listed in the references section of this document.

Key differences are summarized in Table 1-1 to illustrate how the methods and concepts presented in this document supplant pre-GUM techniques. The methods and concepts presented in this document are intended to provide necessary clarification about the topics introduced thus far, as well as other uncertainty analysis issues.

1.4   Application

The established best practices, procedures and illustrative examples contained in this document provide a comprehensive resource for all technical personnel responsible for estimating and reporting measurement uncertainty.

Table 1-1. Comparison of Pre-GUM and GUM Methodologies

Topic	Pre-GUM	GUM
Measurement Error	Measurement errors are categorized as either random or systematic.[3] In this context, random error is defined as the portion of the total measurement error that varies in the short-term when the measurement is repeated. Systematic error is defined as the portion of the total measurement error that remains constant in repeat measurements of a quantity.	The GUM refers only to errors that can occur in a given measurement process and does not differentiate them as random or systematic. Measurement process errors can include repeatability, operator bias, instrument parameter bias, resolution error, errors arising from environmental conditions, or other sources. Additionally, each measurement error, regardless of its origin, is considered to be a random variable that can be characterized by a probability distribution.
Measurement Uncertainty	Many pre-GUM references propose that the uncertainty due to random error be computed by multiplying the standard deviation of a sample of measured values by the Student's t-statistic[4] with 95% confidence level, $t_{95,\nu}$. $$u_{x_{ran}} = t_{95,\nu} s_x \text{ or } u_{\overline{x}_{ran}} = t_{95,\nu} s_{\overline{x}}.$$ The standard deviation, s_x, of a sample of data is $$s_x = \sqrt{\sum_{k=1}^{n} \frac{(x_k - \overline{x})^2}{\nu}}$$ and the standard deviation in the mean value $s_{\overline{x}}$ is $$s_{\overline{x}} = \frac{s_x}{\sqrt{n}}$$	The GUM supplants systematic and random uncertainties with standard uncertainty,[5] which is a statistical quantity equivalent to the standard deviation of the error distribution. By definition, the standard deviation is the square-root of the distribution variance.[6] Therefore, the uncertainty, u_x, in a measurement, $x = x_{true} + \varepsilon_x$, is $$u_x = \sqrt{\text{var}(x)} = \sqrt{\text{var}(x_{true} + \varepsilon_x)}$$ $$= \sqrt{\text{var}(\varepsilon_x)}.$$ In the above equation, x is the measured value, x_{true} is the unknown true value of the measurand at the time of measurement, ε_x is the measurement error and var($\cdot$) is the variance operator. In this regard, uncertainty is not considered to be a $\pm$ limit or interval. The standard uncertainty of a measurement error is determined from Type A or Type B estimates.

---

[3] In the pre-GUM context, the terms random and precision are often used interchangeably, as are the terms systematic and bias.

[4] The Student's t-statistic and confidence level are discussed further in section 2.6.1.

[5] In this document, the terms standard uncertainty and uncertainty are used interchangeably.

[6] A mathematical definition of the distribution variance is presented in section 2.4.

Topic	Pre-GUM	GUM
	where n is the sample size, x_k is the kth measured value, $\bar{x}$ is the sample mean value and ν is the degrees of freedom, equal to n-1.	

Topic	Pre-GUM	GUM
Measurement Uncertainty (continued)	These uncertainties, often expressed by the symbol U_{95}, are more reflective of confidence limits or expanded uncertainties.[7] Pre-GUM references also state that the uncertainty due to systematic error or bias is expressed as $$u_{bias} = \pm B$$ where B is based on past experience, manufacturer specifications, or other information. This uncertainty is also more reflective of confidence limits or an expanded uncertainty.	Type A uncertainty estimates are obtained by the statistical analysis of a sample of measurements. Type B uncertainty estimates are obtained by heuristic means such as past experience, manufacturer specifications, or other information.
Combined Uncertainty	Combining random and systematic uncertainties has been a major issue, often subject to heated debate. The view supported by many data analysts and engineers was to simply add the uncertainties linearly (ADD). $$u_{ADD} = B + t_{95}\frac{s_x}{\sqrt{n}}$$ The view supported by statisticians and measurement science professionals was to combine them in root sum square (RSS). $$u_{RSS} = \sqrt{B^2 + \left(t_{95}\frac{s_x}{\sqrt{n}}\right)^2}$$	Since elemental uncertainties are equal to the square-root of the distribution variance, the variance addition rule is used to combine uncertainties from different error sources. To illustrate the variance addition rule, consider the measurement of a quantity x that involves two error sources ε_1 and ε_2. $$x = x_{true} + \varepsilon_1 + \varepsilon_2$$ The uncertainty in x is obtained from $$u_x = \sqrt{var(x_{true} + \varepsilon_1 + \varepsilon_2)} = \sqrt{var(\varepsilon_1 + \varepsilon_2)}$$ $$= \sqrt{var(\varepsilon_1) + var(\varepsilon_2) + 2\,cov(\varepsilon_1, \varepsilon_2)}$$

---

[7] Confidence limits and expanded uncertainty are also discussed in section 2.6.1.

Topic	Pre-GUM	GUM
	A compromise was eventually proposed[8] in which either method could be used as long as the following constraints were met:	where the covariance term, cov(ε_1, ε_2), is the expected value of the product of the deviations of ε_1 and ε_2 from their respective means. The covariance of two *independent* variables is zero. The covariance can be replaced with

Topic	Pre-GUM	GUM
Combined Uncertainty (continued)	a. The elemental random uncertainties and the elemental systematic uncertainties are combined separately. b. The total random uncertainty and total systematic uncertainty be reported separately. c. The method used to combine the total random and total systematic uncertainties are stated. Ironically, it was also recommended that the RSS method be used to combine the elemental random uncertainties, s_i, and the elemental systematic uncertainties, B_i. $$s = \frac{1}{n}\left[\sum_{i=1}^{K} s_i^2\right]^{1/2} \qquad B = \left[\sum_{i=1}^{K} B_i^2\right]^{1/2}$$ After publication of the GUM, most uncertainty analysis references state that the total random and total systematic uncertainties also be combined in RSS. In many instances, the Student's t-statistic, t_{95}, is set equal to 2 and u_{RSS} is computed to be	$$\rho_{1,2} u_1 u_2 = \text{cov}(\varepsilon_1, \varepsilon_2)$$ where $\rho_{1,2}$ is the correlation coefficient for ε_1 and ε_2 and $$u_1 = \sqrt{\text{var}(\varepsilon_1)} \qquad u_2 = \sqrt{\text{var}(\varepsilon_2)}.$$ Therefore, the uncertainty in x can be expressed as $$u_x = \sqrt{u_1^2 + u_2^2 + 2\rho_{1,2} u_1 u_2}.$$ Since correlation coefficients range from minus one to plus one, this expression provides a more general, mathematically rigorous method for combining uncertainties. For example, if $\rho_{1,2} = 0$ (i.e., statistically independent errors), then the uncertainties are combined using RSS. If $\rho_{1,2} = 1$, then the uncertainties are added. If $\rho_{1,2} = -1$, then the uncertainties are subtracted.

---

[8] Abernathy, R. B. and Ringhauser, B.: "The History and Statistical Development of the New ASME-SAE-AIAA-ISO Measurement Uncertainty Methodology," AIAA/SAE/ASME/ASEE Propulsion Conference, 1985.

Topic	Pre-GUM	GUM
	$$u_{RSS} = \sqrt{B^2 + \left(2\,\frac{s}{\sqrt{n}}\right)^2}$$ Unfortunately, this consensus approach does not eliminate the problems associated with using expanded uncertainties or multiples of standard deviations.	
Degrees of Freedom	Prior to the GUM, there was no way to estimate the degrees of freedom for uncertainties due to systematic error. Consequently, there was no way to compute the degrees of freedom for the combined uncertainty.	Equation G.3 of the GUM $$\nu \approx \frac{1}{2}\frac{u^2(x)}{\sigma^2[u(x)]} \approx \frac{1}{2}\left[\frac{\Delta u(x)}{u(x)}\right]^{-2}$$

Topic	Pre-GUM	GUM
Degrees of Freedom (continued)		provides a relationship for computing the degrees of freedom for a Type B uncertainty estimate where $\sigma^2[u(x)]$ is the variance in the uncertainty estimate, $u(x)$, and $\Delta u(x)$ is the uncertainty in the uncertainty estimate.[9]

Since publication of the GUM, a methodology for determining $\sigma^2[u(x)]$ and computing the degrees of freedom for Type B estimates has been developed.[10]

When uncertainties are combined, it is important to estimate the degrees of freedom for the total uncertainty. The GUM utilizes the Welch-Satterthwaite formula to estimate the effective degrees of freedom, ν_{eff}, for the combined uncertainty. |

---

[9] This equation assumes that the underlying error distribution is normal.

[10] Castrup, H.: "Estimating Category B Degrees of Freedom," presented at the 2000 Measurement Science Conference, January 21, 2001. See also Appendix D.

Topic	Pre-GUM	GUM
		$$v_{eff} = \frac{u_{T*}^4}{\sum\limits_{i=1}^{n} \dfrac{a_i^4 u_i^4}{v_i}}$$ In the above equation, u_i and v_i are the uncertainties and associated degrees of freedom for n error sources, a_i are sensitivity coefficients and the combined or total uncertainty u_{T*} is computed assuming no error source correlations. $$u_{T*} = \sqrt{\sum_{i=1}^{n} a_i^2 u_i^2}$$
Confidence Limits	In pre-GUM references, U_{95} is employed as an equivalent 95% confidence limit $$\bar{x} - U_{95} \leq true\ value \leq \bar{x} + U_{95}$$	The combined or total uncertainty, u_T, and degrees of freedom, v_{eff}, can be used to establish the upper and lower limits that contain the true value (estimated by the mean value $\bar{x}$), with some specified confidence level, p. Confidence limits are expressed as

Topic	Pre-GUM	GUM
Confidence Limits (continued)	where $$U_{95} = \sqrt{B^2 + \left(t_{95}\, \frac{s}{\sqrt{n}}\right)^2}$$	$$\bar{x} - t_{\alpha/2, v_{eff}}\, u_T \leq true\ value \leq \bar{x} + t_{\alpha/2, v_{eff}}\, u_T$$ where $\alpha = 1 - p$ and the t-statistic, $t_{\alpha/2 veff}$, is a function of both the degrees of freedom and the confidence level. The GUM introduces an expanded uncertainty, ku, as an approximate confidence limit, in which a coverage factor k is used. $$\bar{x} - ku_T \leq true\ value \leq \bar{x} + ku_T$$ In most cases, a value of $k = 2$ is used to approximate a 95% confidence level for normally distributed errors.

Topic	Pre-GUM	GUM
		To be useful in managing errors, the coverage factor should be based on both a confidence level and the degrees of freedom for the uncertainty estimate. This is achieved with the Student's t-statistic, $t_{\alpha/2,\nu}$.

Confidence limits and expanded uncertainty are discussed further in section 2.6.1. |

CHAPTER 2: BASIC CONCEPTS AND METHODS

A measurement is a process whereby the value of a quantity is estimated. All measurements are accompanied by error.[11] Our lack of knowledge about the sign and magnitude of measurement error is called measurement uncertainty. A measurement uncertainty estimate is the characterization of what we know statistically about the measurement error. Therefore, a measurement result is only complete when accompanied by a statement of the uncertainty in that result.

This chapter describes the basic concepts and methods used to estimate measurement uncertainty.[12] The general uncertainty analysis procedure consists of the following steps:

1. Define the Measurement Process
2. Develop the Error Model
3. Identify the Error Sources and Distributions
4. Estimate Uncertainties
5. Combine Uncertainties
6. Report the Analysis Results

The following sections discuss these analysis steps and clarify the relationship between measurement error and uncertainty. A discussion on using uncertainty estimates to compute confidence intervals and expanded uncertainties is also included.

2.1   Define the Measurement Process

The first step in any uncertainty analysis is to identify the physical quantity that is measured. This quantity, sometimes referred to as the "measurand," may be a directly measured value or derived from the measurement of other quantities. The former type of measurements are called "direct measurements," while the latter are called "multivariate measurements."

For multivariate measurements, it is important to develop an equation that defines the mathematical relationship between the derived quantity of interest and the measured quantities. For a case involving three measured quantities x, y, and z, this equation can be written

$$q = f\left(x, y, z\right) \qquad\qquad (2\text{-}1)$$

where

$\quad q \quad = \quad$ quantity of interest

$\quad f \quad = \quad$ mathematical function that relates q to measured quantities x, y, and z.

At this initial stage of the analysis, it is also important to describe the test setup, environmental conditions, technical information about the instruments, reference standards, or other equipment used and the procedure for obtaining the measurement(s). This information will be used to identify measurement process errors and estimate uncertainties.

---

[11] The relationship between a measured quantity and measurement error is defined in section 2.2.

[12] The methodology of the GUM is employed throughout this document. The same applies to specific procedures and techniques unless otherwise indicated.

2.2   Develop the Error Model

An error model is an algebraic expression that defines the total error in the value of a quantity in terms of all relevant measurement process or component errors. The error model for the quantity q defined in equation (2-1) is

$$\varepsilon_q = c_x \varepsilon_x + c_y \varepsilon_y + c_z \varepsilon_z \qquad (2\text{-}2)$$

where

ε_q = error in q
ε_x = error in the measured quantity x
ε_y = error in the measured quantity y
ε_z = error in the measured quantity z

and c_x, c_y and c_z are sensitivity coefficients that determine the relative contribution of the errors in x, y and z to the total error in q. The sensitivity coefficients are defined below.[13]

$$c_x = \left(\frac{\partial q}{\partial x} \right) \, , \; c_y = \left(\frac{\partial q}{\partial y} \right) \, , \; c_z = \left(\frac{\partial q}{\partial z} \right)$$

In any given measurement scenario, each measured quantity is also accompanied by measurement error. The basic relationship between the measured quantity x and the measurement error ε_x is given in equation (2-3).

$$x = x_{\text{true}} + \varepsilon_x \qquad (2\text{-}3)$$

The error model for ε_x is the sum of the errors encountered during the measurement process and is expressed as

$$\varepsilon_x = \varepsilon_1 + \varepsilon_2 + \cdots + \varepsilon_k \qquad (2\text{-}4)$$

where the numbered subscripts signify the different measurement process errors.

2.3   Identify Measurement Errors and Distributions

Measurement process errors are the basic elements of uncertainty analysis. Once these fundamental error sources have been identified, then uncertainty estimates for these errors can be developed.

The errors most often encountered in making measurements include, but are not limited to the following:

- Reference Attribute Bias
- Repeatability
- Resolution Error
- Operator Bias
- Environmental Factors Error
- Computation Error

---

[13] Detailed analysis procedures for multivariate measurements are presented in Chapter 6.

Reference Attribute Bias

Calibrations are performed to obtain an estimate of the value or bias of selected unit-under-test (UUT) attributes by comparison to corresponding measurement reference attributes. The error in the value of a reference attribute, at any instant in time, is composed of a systematic component and a random component. Reference attribute bias is the systematic error component that persists from measurement to measurement during a measurement session.[14] Attribute bias excludes resolution error, random error, operator bias and other error sources that are not properties of the attribute.

Repeatability

Repeatability is a random error that manifests itself as differences in measured value from measurement to measurement during a measurement session. It is important to note that, random variations in a measured quantity or UUT attribute are not separable from random variations in the reference attribute or random variations due to other error sources.

Resolution Error

Reference attributes and/or UUT attributes may provide indications of sensed or stimulated values with some finite precision. The smallest discernible value indicated in a measurement comprises the resolution of the measurement. For example, a voltmeter may indicate values to four, five or six significant digits. A tape measure may provide length indications in meters, centimeters or millimeters. A scale may indicate weight in terms of kg, g, mg or μg.

The basic error model for resolution error, ε_{res}, is

$$\varepsilon_{res} = x_{indicated} - x_{sensed}$$

where x_{sensed} is a "measured" value detected by a sensor or provided by a stimulus and $x_{indicated}$ is the indicated representation of x_{sensed}.

Operator Bias

Errors can be introduced by the person or operator making the measurement. Because of the potential for human operators to acquire measurement information from an individual perspective or to produce a systematic bias in a measurement result, it sometimes happens that two operators observing the same measurement result will systematically perceive or produce different measured values.

In reality, operator bias has a somewhat random character due to inconsistencies in human behavior and response. The random contribution is included in measurement repeatability and the systematic contribution is the operator bias.

Environmental Factors Error

Errors can result from variations in environmental conditions, such as temperature, vibration, humidity or stray emf. Additional errors are introduced when measurement results are corrected for environmental conditions. For example, when correcting a length measurement for thermal

---

[14] A measurement session is considered to be an activity in which a measurement or sample of measurements is taken under fixed conditions, usually for a period of time measured in seconds, minutes or, at most, hours.

expansion, the error in the temperature measurement will introduce an error in the length correction. The uncertainty in the correction error is a function of the uncertainty in the error in the environmental factor.[15]

Computation Error

Data processing errors result from computation round-off or truncation, numerical interpolation of tabulated values, or the use of curve fit equations. For example, in the regression analysis of a range of values, the standard error of estimate quantifies the difference between the measured values and the values estimated from the regression equation.[16]

A regression analysis that has a small standard error of estimate has data points that are very close to the regression line. Conversely, a large standard error of estimate results when data points are widely dispersed around the regression line. However, if another sample of data were collected, then a different regression line would result. The standard error of the forecast accounts for the dispersion of various regression lines that would be generated from multiple sample sets around the true population regression line. The standard error of forecast is a function of the standard error of estimate and the measured value and should be used when estimating uncertainty due to regression error.

Repeatability and Resolution Error

In some measurement situations, repeatability may be considered to be a manifestation of resolution error. The following cases should be considered when determining whether or not to include repeatability and resolution as separate error sources.

Case 1 – Values obtained in a random sample of measurements exhibit just two values and the difference between these values is equal to the smallest increment of resolution. In this case, it can be concluded that "background noise" random variations are occurring that are beyond the resolution of the measurement. Consequently, repeatability cannot be identified as a separate error source because the apparent random variations are due to resolution error. Accordingly, the uncertainty due to resolution error should be included in the total measurement uncertainty but the uncertainty due to repeatability should not be included.

Case 2 – Values obtained in a random sample vary in magnitude substantially greater than the smallest increment of resolution. In this case, repeatability cannot be ignored as an error source. In addition, since each sampled value is subject to resolution error, it should also be included. Accordingly, the total measurement uncertainty must include contributions from both repeatability uncertainty and resolution uncertainty.

Case 3 – Values obtained in a random sample of measurements vary in magnitude somewhat greater than the smallest increment of resolution but not substantially greater. In this case, error due to repeatability is partly separable from resolution error, but it becomes a matter of opinion as to whether to include repeatability and resolution error in the total measurement error. Until a clear solution to the problem is found, it is best to include both repeatability and resolution error.

In summary, if measurement repeatability is smaller than the display resolution, only resolution

---

[15] In the length correction scenario, error in the coefficient of thermal expansion may also need to be taken into account.

[16] Hanke, J. et al.: *Statistical Decision Models for Management*, Allyn and Bacon, Inc. 1984.

error should be included in the uncertainty analysis. If measurement repeatability is larger than the display resolution, then both error sources should be included in the uncertainty analysis.

2.3.1 Error Distributions

Recall from the GUM methodology discussed in Chapter 1, that measurement uncertainty is the square root of the variance of the error distribution.[17] To better understand the relationship between measurement error and measurement uncertainty, measurement error distributions must be discussed in some detail.

An important aspect of the uncertainty analysis process is the fact that measurement errors can be characterized by probability distributions. This is stated in Axiom 1.

Axiom 1 - Measurement errors are random variables that follow probability distributions.

The probability distribution for a type of measurement error is a mathematical description that relates the frequency of occurrence of values to the values themselves. Error distributions include, but are not limited to normal, lognormal, uniform (rectangular), triangular, quadratic, cosine, exponential, U-shaped and trapezoidal.

Each distribution is characterized by a set of statistics. The statistics most often used in uncertainty analysis are the mean or mode and the standard deviation. With the lognormal distribution, a limiting value and the median value are also used. Probability distributions used in measurement applications are described in Appendix B. Probability density functions for selected distributions are summarized in Table 2-1.

Table 2-1. Probability Distributions

Distribution	Distribution Plot	Probability Density Function
Normal		$$f(\varepsilon) = \frac{1}{\sqrt{2\pi}\sigma} e^{-(\varepsilon)^2/2\sigma^2}$$ where σ is the standard deviation of the distribution.
Lognormal		$$f(\varepsilon) = \frac{1}{\sqrt{2\pi}\lambda|\varepsilon - q|} \exp\left\{ -\frac{\left[\ln\left(\frac{\varepsilon - q}{(m - q)}\right)\right]^2}{2\lambda^2} \right\}$$ where q is a physical limit for ε, m is the distribution median, and λ is a shape parameter.

---

[17] The basis for the mathematical relationship between error and uncertainty is presented in section 2.4.

13

Distribution	Distribution Plot	Probability Density Function
Quadratic		$$f(\varepsilon) = \begin{cases} \dfrac{3}{4a}\left[1-(\varepsilon/a)^2\right], & -a \le \varepsilon \le a \\ 0, & \text{otherwise} \end{cases}$$ where $\pm a$ are the minimum distribution bounding limits.
Cosine		$$f(\varepsilon) = \begin{cases} \dfrac{1}{2a}\left[1+\cos(\pi\varepsilon/a)\right], & -a \le \varepsilon \le a \\ 0, & \text{otherwise} \end{cases}$$ where $\pm a$ are the minimum distribution bounding limits.
Uniform (Rectangular)		$$f(\varepsilon) = \begin{cases} \dfrac{1}{2a}, & -a \le \varepsilon \le a \\ 0, & \text{otherwise} \end{cases}$$ where $\pm a$ are the minimum distribution bounding limits.
Triangular		$$f(\varepsilon) = \begin{cases} (\varepsilon+a)/a^2, & -a \le \varepsilon \le 0 \\ (a-\varepsilon)/a^2, & 0 \le \varepsilon \le a \\ 0, & \text{otherwise} \end{cases}$$ where $\pm a$ are the minimum distribution bounding limits.
Trapezoidal		$$f(\varepsilon) = \begin{cases} \dfrac{1}{d^2+c^2}(d+\varepsilon), & -d \le \varepsilon \le -c \\ \dfrac{1}{d+c}, & -c \le \varepsilon \le c \\ 0, & \text{otherwise} \end{cases}$$ where $\pm d$ are the minimum distribution bounding limits.
U-Shaped		$$f(\varepsilon) = \begin{cases} \dfrac{1}{\pi\sqrt{a^2+\varepsilon^2}}, & -a \le \varepsilon \le a \\ 0, & \text{otherwise} \end{cases}$$ where $\pm a$ are the minimum distribution bounding limits.

2.3.2 Choosing the Appropriate Distribution

The normal and lognormal distributions are relevant to most real world measurement applications. Other distributions are also possible, such as the uniform, triangular, quadratic, cosine, exponential, and U-shaped, although they have limited applicability. Some recommendations for selecting the appropriate distribution for a particular measurement error source are as follows:

a. The normal distribution should be applied as the default distribution, unless information to the contrary is available.

b. Apply the lognormal distribution if it is suspected that the distribution of the value of interest is skewed (i.e., non-symmetric) and bounded on one side.[18]

c. If 100% containment has been observed and minimum bounding limits are known, then the following is recommended:

   i. Apply the cosine distribution if the value of interest has been subjected to random usage or handling stress, and is assumed to possess a central tendency.

   ii. Apply the quadratic distribution if it is suspected that values are more evenly distributed.

   iii. The triangular distribution may be applicable, under certain circumstances, when dealing with parameters following testing or calibration. It is also the distribution of the sum of two uniformly distributed errors with equal means and bounding limits.

   iv. The U-shaped distribution is applicable to quantities controlled by feedback from sensed values, such as automated environmental control systems.

   v. Apply the uniform distribution if the value of interest is the resolution uncertainty of a digital readout. This distribution is also applicable to estimating the uncertainty due to quantization error and the uncertainty in RF phase angle.

More specific criteria for correctly selecting the uniform distribution and example cases that satisfy this criteria are given in Appendix B.

2.4 Estimate Uncertainties

As previously discussed, an error distribution tells us whether an error or a range of errors is likely or unlikely to occur. It provides a mathematical description of how likely we are to experience (measure) certain values. With a basic understanding of error distributions and their statistics, we can estimate uncertainties.[19] We begin with the statistical quantity called the **variance**.

2.4.1 Distribution Variance

Variance is defined as the mean square dispersion of the distribution about its mean value.

---

[18] In using the normal or lognormal distribution, some effort must be made to estimate a containment probability. This is discussed in more detail in Chapter 3.

[19] To ensure validity, the distribution selected to estimate uncertainty for a given error source should provide the most realistic statistical characteristics.

$$\text{var}(x) = \text{Mean Square Dispersion in } x.$$

If a variable x follows a probability distribution, described by a probability density function $f(x)$, then the mean square dispersion or variance of the distribution is given by

$$\text{var}(x) = \int_{-\infty}^{\infty}(x - \mu_x)^2 f(x)dx \qquad (2\text{-}5)$$

where μ_x is the mean of x, sometimes referred to as the *expectation value* for x. In speaking of variations in x that are the result of measurement error, we take μ_x to be the true value of the quantity being measured. From equation (2-3), we can write $\varepsilon_x = x - \mu_x$, and equation (2-5) can be expressed as

$$\begin{aligned}\text{var}(x) &= \int_{-\infty}^{\infty} \varepsilon_x^2 f\left[x(\varepsilon_x)\right]d\varepsilon_x \\ &= \text{var}(\varepsilon_x)\end{aligned} \qquad (2\text{-}6)$$

where $x(\varepsilon_x) = \varepsilon_x + \mu_x$. Because of the form of this definition, the variance is also referred to as the mean square error.

Equation (2-6) shows that, if a quantity x is a random variable representing a population of measurements, then the variance in x is just the variance in the error in x

$$\text{var}(x) = \text{var}(x_{\text{true}} + \varepsilon_x) = \text{var}(\varepsilon_x). \qquad (2\text{-}7)$$

By definition, the standard deviation is the square root of the distribution variance or mean square error. The uncertainty in a measurement quantity is equivalent to the standard deviation of the error distribution. This leads to Axiom 2.

> **Axiom 2 -** **The uncertainty in a measurement is the square root of the variance in the measurement error.**

Axiom 2 provides the crucial link between measurement error and measurement uncertainty. If x is a measured value, then

$$\begin{aligned}u_x &= \sqrt{\text{var}(x)} \\ &= \sqrt{\text{var}(\varepsilon_x)} \\ &= u_{\varepsilon_x}.\end{aligned} \qquad (2\text{-}8)$$

Equation (2-8) provides a third axiom that forms a solid and productive basis upon which uncertainties can be estimated.

> **Axiom 3 -** **The uncertainty in a measured value is equal to the uncertainty in the measurement error.**

Axiom 3, together with Axioms 1 and 2, allows the computation and combination of measurement uncertainty to be rigorously carried out by

1. drawing attention to what it is that we are uncertain of in making measurements, and

2. allowing for the development of measurement uncertainty models for measurement scenarios of any complexity.

There are two approaches to estimating the variance of an error distribution and, thus, the uncertainty in the measurement error. Type A estimates involve data sampling and analysis. Type B estimates use technical knowledge or recollected experience of measurement processes. The basic methods used to make Type A and Type B uncertainty estimates are presented in Chapter 3.

2.5 Combine Uncertainties

Axiom 2 states that the uncertainty in the value of an error is equal to the square root of the variance of the error distribution. As a consequence, we can apply the variance addition rule to obtain a method for correctly combining uncertainties from different error sources.

2.5.1 Variance Addition Rule – Direct Measurements

For purposes of illustration, consider a measured quantity $x = x_{true} + \varepsilon_x$. We know that the total error, ε_x, consists of measurement process errors

$$\varepsilon_x = \varepsilon_1 + \varepsilon_2 + \cdots + \varepsilon_k = \sum_{i=1}^{k} \varepsilon_i$$

where ε_i represents the ith error source and k is the total number of errors.

Applying the variance addition rule to ε_x yields

$$
\begin{aligned}
\text{var}(\varepsilon_x) &= \text{var}(\varepsilon_1 + \varepsilon_2 + \cdots + \varepsilon_k) \\
&= \text{var}(\varepsilon_1) + \text{var}(\varepsilon_2) + \cdots + \text{var}(\varepsilon_k) + 2\text{cov}(\varepsilon_1, \varepsilon_2) + \\
&\quad 2\text{cov}(\varepsilon_1, \varepsilon_3) + \cdots + 2\text{cov}(\varepsilon_{k-2}, \varepsilon_k) + 2\text{cov}(\varepsilon_{k-1}, \varepsilon_k) \\
&= \sum_{i=1}^{k} \text{var}(\varepsilon_i) + 2\sum_{i=1}^{k-1} \sum_{j=i+1}^{k} \text{cov}(\varepsilon_i, \varepsilon_j)
\end{aligned}
\tag{2-9}
$$

where $\text{cov}(\varepsilon_i, \varepsilon_j)$ is the **covariance** between measurement process errors. Covariance is defined in Section 2.5.3.

2.5.2 The Variance Addition - General Model

Now consider a more general case of the variance addition rule. For illustration, consider a quantity z defined from the following equation

$$z = ax + by$$

where x and y are measured quantities and the coefficients a and b are constants. Using equation

(2-3),

$$z = a(x_{true} + \varepsilon_x) + b(y_{true} + \varepsilon_y)$$
$$= ax_{true} + by_{true} + a\varepsilon_x + b\varepsilon_y$$
$$= z_{true} + \varepsilon_z$$

where

$$z_{true} = ax_{true} + by_{true}$$

and

$$\varepsilon_z = a\varepsilon_x + b\varepsilon_y.$$

The variance of ε_z is expressed as

$$\text{var}(\varepsilon_z) = \text{var}(a\varepsilon_x + b\varepsilon_y)$$
$$= a^2 \, \text{var}(\varepsilon_x) + b^2 \, \text{var}(\varepsilon_y) + 2ab \, \text{cov}(\varepsilon_x, \varepsilon_y)$$

where the last term is the covariance between ε_x and ε_y. From Axiom 2, $\text{var}(\varepsilon_z)$ is expressed as

$$\text{var}\left(\varepsilon_z\right) = u_{\varepsilon_z}^2$$
$$= a^2 u_{\varepsilon_x}^2 + b^2 u_{\varepsilon_y}^2 + 2ab\,\text{cov}(\varepsilon_x,\varepsilon_y) \tag{2-10}$$

where u_{ε_x} and u_{ε_y} are the uncertainties in ε_x and ε_y, respectively.

2.5.3 Error Correlations – Direct Measurements

If two variables ε_1 and ε_2 are described by a joint probability density function $f(\varepsilon_1, \varepsilon_2)$, then the covariance of ε_1 and ε_2 is given by

$$\text{cov}(\varepsilon_1,\varepsilon_2) = \int_{-\infty}^{\infty} d\varepsilon_1 \int_{-\infty}^{\infty} \varepsilon_1\varepsilon_2 f(\varepsilon_1,\varepsilon_2) d\varepsilon_2. \tag{2-11}$$

The covariance of two random variables is a statistical assessment of their mutual dependence. Because covariances can have inconvenient physical dimensions, they are rarely used explicitly. Instead, we use the **correlation coefficient**, $\rho_{\varepsilon_i\varepsilon_j}$, which is defined as

$$\rho_{\varepsilon_i\varepsilon_j} = \frac{\text{cov}(\varepsilon_i,\varepsilon_j)}{\sqrt{\text{var}(\varepsilon_i)\,\text{var}(\varepsilon_j)}} = \frac{\text{cov}(\varepsilon_i,\varepsilon_j)}{u_{\varepsilon_i}u_{\varepsilon_j}} \tag{2-12}$$

where u_{ε_i} and u_{ε_j} are the ith and jth measurement process uncertainties. The correlation coefficient provides an assessment of the relative mutual dependence of two random variables. The correlation coefficient is a dimensionless number ranging in value from -1 to 1.

If we recall Axiom 2, equation (2-9) can be expressed as

$$u_{\varepsilon_x}^2 = \sum_{i=1}^{k} \text{var}(\varepsilon_i) + 2 \sum_{i=1}^{k-1} \sum_{j=i+1}^{k} \rho_{\varepsilon_i \varepsilon_j} u_{\varepsilon_i} u_{\varepsilon_j}$$

$$= \sum_{i=1}^{k} u_{\varepsilon_i}^2 + 2 \sum_{i=1}^{k-1} \sum_{j=i+1}^{k} \rho_{\varepsilon_i \varepsilon_j} u_{\varepsilon_i} u_{\varepsilon_j} \qquad (2\text{-}13)$$

2.5.4 Cross-Correlations between Error Components

From equations (2-12) and (2-13), the correlation coefficient for ε_x and ε_y is

$$\rho_{\varepsilon_x \varepsilon_y} = \frac{\text{cov}(\varepsilon_x, \varepsilon_y)}{u_{\varepsilon_x} u_{\varepsilon_y}} \qquad (2\text{-}14)$$

and equation (2-10) becomes

$$u_{\varepsilon_z}^2 = a^2 u_{\varepsilon_x}^2 + b^2 u_{\varepsilon_y}^2 + 2ab\rho_{\varepsilon_x \varepsilon_y} u_{\varepsilon_x} u_{\varepsilon_y}. \qquad (2\text{-}15)$$

Equation (2-15) can be generalized to cases where there are k measured quantities and corresponding error components $\varepsilon_1, \varepsilon_2, ..., \varepsilon_k$ for these quantities.

$$\text{var}(\sum_{r=1}^{k} a_r \varepsilon_r) = \sum_{r=1}^{k} a_r^2 \, \text{var}(\varepsilon_r) + 2 \sum_{r=1}^{k-1} \sum_{q=r+1}^{k} a_r a_q \rho_{\varepsilon_r \varepsilon_q} u_{\varepsilon_r} u_{\varepsilon_q}$$

$$= \sum_{r=1}^{k} a_r^2 u_{\varepsilon_r}^2 + 2 \sum_{r=1}^{k-1} \sum_{q=r+1}^{k} a_r a_q \rho_{\varepsilon_r \varepsilon_q} u_{\varepsilon_r} u_{\varepsilon_q} \qquad (2\text{-}16)$$

where u_{ε_r} and u_{ε_q} are the total uncertainties for the rth and qth error components, respectively and ρ_{rq} is the cross-correlation between these error components.

Each error component is comprised of measurement process errors, such as measurement reference bias, repeatability, resolution error, etc. Hence, we decompose ε_r as

$$\varepsilon_r = \varepsilon_{r,1} + \varepsilon_{r,2} + ... + \varepsilon_{r,l} \qquad (2\text{-}17)$$

where l denotes the number of measurement process errors.

The cross-correlation coefficient between measurement process errors for the error components $\varepsilon_{r,i}$ and $\varepsilon_{q,j}$, is denoted by $\rho_{\varepsilon_{r,i} \varepsilon_{q,j}}$ and written

$$\rho_{\varepsilon_{r,i} \varepsilon_{q,j}} = \frac{\text{cov}\left(\varepsilon_{r,i}, \varepsilon_{q,j}\right)}{u_{\varepsilon_{r,i}} u_{\varepsilon_{q,j}}} \qquad (2\text{-}18)$$

where $u_{\varepsilon_{r,i}}$ and $u_{\varepsilon_{q,j}}$ are the measurement process uncertainties for the rth and qth error components, respectively, and

$$u_{\varepsilon_{r,i}} = \sqrt{\mathrm{var}(\varepsilon_{r,i})}.$$ (2-19)

Returning to equation (2-16), the correlation coefficient for u_{ε_r} and u_{ε_q} is

$$\rho_{\varepsilon_r \varepsilon_q} = \frac{1}{u_{\varepsilon_r} u_{\varepsilon_q}} \sum_{i=1}^{l} \sum_{j=1}^{m} \rho_{\varepsilon_{r,i} \varepsilon_{q,j}} u_{\varepsilon_{r,i}} u_{\varepsilon_{q,j}}$$ (2-20)

where the total number of process uncertainties for the rth and qth measured quantities are l and m, respectively. As equation (2-20) shows, the correlation coefficient, ρ_{rqij}, accounts for cross-correlations between measurement process uncertainties for the rth and qth error components.

2.5.5 Combined Uncertainty

The variance addition rule provides a logical approach for computing the overall, combined uncertainty that accounts for correlations between error sources. Given equations (2-16) and (2-20), the total uncertainty, u_T, can be generally expressed as

$$u_T = \sqrt{ \sum_{r=1}^{k} a_r^2 u_{\varepsilon_r}^2 + 2 \sum_{r=1}^{k-1} \sum_{q=r+1}^{k} a_r a_q \sum_{i=1}^{l} \sum_{j=1}^{m} \rho_{\varepsilon_{r,i} \varepsilon_{q,j}} u_{\varepsilon_{r,i}} u_{\varepsilon_{q,j}} }$$ (2-21)

From the above equation, one can surmise that uncertainties are not always combined using the root sum square (RSS) method.

2.5.6 Establishing Correlations

To assess the impact of correlated errors on the combined uncertainty, consider the measurement of a quantity x that involves two error sources ε_1 and ε_2

$$x = x_{true} + \varepsilon_1 + \varepsilon_2.$$

From Axioms 2 and 3 and the variance addition rule, the uncertainty in x is obtained from

$$u_x = \sqrt{\mathrm{var}(x_{true} + \varepsilon_1 + \varepsilon_2)} = \sqrt{\mathrm{var}(\varepsilon_1 + \varepsilon_2)}$$

$$= \sqrt{u_1^2 + u_2^2 + 2\rho_{1,2} u_1 u_2}.$$

The correlation coefficient, $\rho_{1,2}$, for two error sources can range in value from -1 to +1.

Statistically Independent Error Sources

If the two error sources are statistically independent, then $\rho_{1,2} = 0$ and $u_x = \sqrt{u_1^2 + u_2^2}$.

Therefore, uncertainties of statistically independent error sources are combined using the RSS method.

20

If $\rho_{1,2} = 1$, then $u_x = \sqrt{u_1^2 + u_2^2 + 2u_1u_2} = \sqrt{(u_1 + u_2)^2} = u_1 + u_2$. Therefore, the uncertainties are combined linearly.

When two error sources are strongly correlated and compensate for one another, then $\rho_{1,2} = -1$ and $u_x = \sqrt{u_1^2 + u_2^2 - 2u_1u_2} = \sqrt{(u_1 - u_2)^2} = |u_1 - u_2|$. Therefore, the combined uncertainty is the absolute value of the difference between the individual uncertainties.

There typically aren't any correlations between measurement process errors for a given quantity. In general, it is safe to assume that there are no correlations between the following measurement process errors:

- Repeatability and Reference Attribute Bias ($\rho_{\text{ran,bias}} = 0$)
- Repeatability and Operator Bias ($\rho_{\text{ran,oper}} = 0$)
- Reference Attribute Bias and Resolution Error ($\rho_{\text{bias,res}} = 0$)
- Reference Attribute Bias and Operator Bias ($\rho_{\text{bias,oper}} = 0$)
- Operator Bias and Environmental Factors Error ($\rho_{\text{oper,env}} = 0$)
- Resolution Error and Environmental Factors Error ($\rho_{\text{res, env}} = 0$)
- Digital Resolution Error and Operator Bias ($\rho_{\text{res, oper}} = 0$)

Cross-Correlations

Instances may arise where measurement process errors for different error components are correlated. In this case, equation (2-20) must be applied to account for cross-correlations between measurement components. Accounting for cross-correlations is discussed further in Chapter 6.

2.5.7   Degrees of Freedom

Generally speaking, degrees of freedom signifies the amount of information or knowledge that went into an uncertainty estimate. Therefore, when uncertainties are combined, we need to know the degrees of freedom for the total uncertainty. Unfortunately, the degrees of freedom for a combined uncertainty estimate is not a simple sum of the degrees of freedom for each uncertainty component.

The effective degrees of freedom, v_{eff}, for the total uncertainty, u_T, resulting from the combination of uncertainties u_i and associated degrees of freedom, v_i, for n error sources can be estimated via the Welch-Satterthwaite formula given in equation (2-22)

$$v_{eff} = \frac{u_{T*}^4}{\sum_{i=1}^{n} \frac{a_i^4 u_i^4}{v_i}} \tag{2-22}$$

where u_{T*} is the total or combined uncertainty computed assuming no error correlations.

$$u_{T*} = \sqrt{\sum_{i=1}^{n} a_i^2 u_i^2}$$ (2-23)

Note: While the Welch-Satterthwaite formula is applicable for statistically independent, normally distributed error sources, it can usually be thought of as a fair approximation in cases where error sources are not statistically independent.

Determining the degrees of freedom for Type A and Type B uncertainty estimates is discussed in Chapter 3.

2.6   Report the Analysis Results

Reporting the results of an uncertainty analysis is an important aspect of measurement quality assurance. Therefore, the analysis results must be reported in a way that can be readily understood and interpreted by others.

Section 7 of the GUM recommends that the following information be included:

1. The estimated value of the quantity of interest (measurand) and its combined uncertainty and degrees of freedom.

2. The functional relationship between the quantity of interest and the measured components, along with the sensitivity coefficients.

3. The value of each measurement component and its combined uncertainty and degrees of freedom

4. A list of the measurement process uncertainties and associated degrees of freedom for each component, along with a description of how they were estimated.

5. A list of applicable correlation coefficients, including any cross-correlations between component uncertainties.

It is also a good practice to provide a brief description of the measurement process, including the procedures and instrumentation used, and additional data, tables and plots that help clarify the analysis results.

When reporting the uncertainty in a measured value, it is often desirable to include confidence limits or expanded uncertainty. Therefore, some discussion about confidence limits and expanded uncertainty is provided in the following section.

2.6.1   Confidence Limits and Expanded Uncertainty

In statistics, we make inferences about population parameters, such as the mean value and standard deviation, through the analysis of sampled data or other heuristic information. Confidence limits provide a numerical interval which contains the population parameter of interest with some probability.[20] Confidence limits are computed using either the normal or Student's t distribution.[21]

---

[20] Hanke, J. et al.: *Statistical Decision Models for Management*, Allyn and Bacon, Inc., 1984.

[21] The Student's t distribution is a symmetric distribution that approaches the normal distribution as the degrees of freedom approach infinity.

In reporting measurement results, the uncertainty, u, and its associated degrees of freedom, v, can be used to establish confidence limits that contain the true value, μ (estimated by a sample mean value $\bar{x}$), with some specified confidence level or probability, p. In this application, the confidence limits are expressed as

$$\bar{x} - t_{\alpha/2,v}\, u \le \mu \le \bar{x} + t_{\alpha/2,v}\, u \qquad\qquad (2\text{-}24)$$

where the multiplier is the t-statistic, $t_{\alpha/2,v}$, and $\alpha = 1 - p$ is the significance level. Values for $t_{\alpha/2,v}$ are obtained from the percentiles of the probability density function for the Student's t distribution.

As seen from equation (2-24), the width of the confidence limits or interval is dependent on three factors:

1. the confidence level
2. the estimated uncertainty
3. the degrees of freedom.

The development and application of confidence limits are discussed further in Chapters 3 and 4.

The GUM defines the term **expanded uncertainty** as "the quantity defining an interval about the result of a measurement that may be expected to encompass a large fraction of the distribution of values that could reasonably be attributed to the measurand."

This means that the expanded uncertainty is basically defined as an interval that is expected to contain the true value of the measurand. In this context, the expanded uncertainty, ku, is offered as approximate confidence limits, in which the coverage factor, k, is used in place of the t-statistic

$$\bar{x} - ku \le x_{true} \le \bar{x} + ku\,. \qquad\qquad (2\text{-}25)$$

The introduction of the expanded uncertainty was meant to clarify the concept of uncertainty, but confusion over and misapplication of this term persisted since the GUM was first released. To mitigate this problem, the GUM also introduced the term "standard uncertainty" to help distinguish uncertainty from expanded uncertainty. However, in practice, the term expanded uncertainty and uncertainty are often used interchangeably. This, of course, can lead to incorrect inferences and miscommunications.

> **Note**: The use of the term uncertainty to represent an expanded uncertainty is not a recommended practice.

The use of coverage factors in lieu of the t-statistic emerged as an artifice to "emulate" confidence limits in cases where the total uncertainty is a Type B estimate or is composed of both Type A and Type B estimates.

Not being a statistical quantity in the purest sense, a Type B estimate was not considered to be associated with definable degrees of freedom that could be regarded as quantifying the amount of information used in producing the estimate. Accordingly, if used alone or combined with a Type A estimate, the result was not viewed as being a true statistic.

As is shown in Appendix D, we now have the means to estimate the degrees of freedom for Type B estimates in such a way that they can be considered on an approximately equal statistical footing with Type A estimates. Consequently, Type B uncertainty estimates can be used to determine confidence limits, conduct statistical tests, evaluate decisions, etc.

CHAPTER 3: ESTIMATING UNCERTAINTY

There are two approaches to estimating measurement uncertainty. Type A uncertainty estimates involve data sampling and analysis. Type B uncertainty estimates use engineering knowledge or recollected experience of measurement processes. This chapter discusses sample statistics used to make Type A uncertainty estimates and heuristic methods used to make Type B uncertainty estimates.

3.1   Type A Estimates

A Type A uncertainty estimate is defined as an estimate obtained from a sample of data. Data sampling involves making repeat measurements of the quantity of interest. It is important that each repeat measurement is independent, representative and taken randomly.

Random sampling is a cornerstone for obtaining relevant statistical information. Therefore, Type A estimates usually apply to the uncertainty due to repeatability or random error. The data used for Type A uncertainty estimates typically consist of sampled values. However, the data may be comprised of sampled mean values or sampled cells. The computed statistics vary slightly depending on the sample type.

Statistical analysis of sampled values will be presented herein for illustration. Statistical analysis methods for all three sample types are presented in Appendix C, along with topics on outlier removal and normality testing.

3.1.1   Statistics for Sampled Values

Because the data sample is drawn from a population[22] of values, we make inferences about the population from certain sample statistics and from assumptions about the way the population of values is distributed. A sample histogram can aid in our attempt to picture the population distribution.

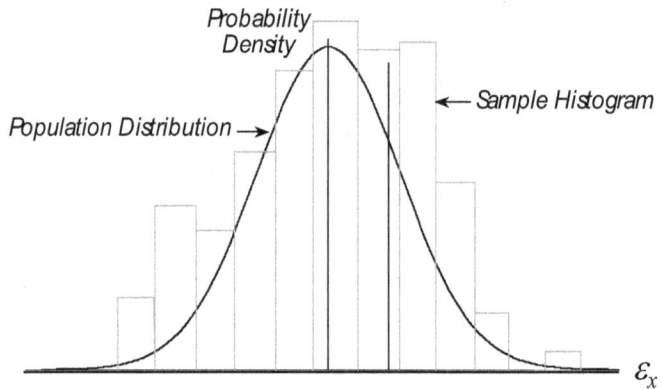

Figure 3-1. Repeatability Distribution

The normal distribution is ordinarily assumed to be the underlying distribution for repeatability or random error. When samples are taken, the sample mean and the sample standard deviation are computed and assumed to represent the mean and standard deviation of the population

---

[22] In statistics, a population is the total set of possible values for a random variable under consideration.

distribution. However, this equivalence is only approximate. To account for this, the Student's *t* distribution is used in place of the normal distribution to compute confidence limits around the sample mean.

The sample mean, $\bar{x}$, is obtained by taking the average of the sampled values. The average value is computed by summing the values sampled and dividing them by the sample size, *n*.

$$\bar{x} = \frac{1}{n}\left(x_1 + x_2 + \ldots + x_n\right) = \frac{1}{n}\sum_{i=1}^{n} x_i \qquad (3\text{-}1)$$

The sample mean can be thought of as an estimate of the value that we expect to get when we make a measurement. This "expectation value" is called the population mean, which is expressed by the symbol μ.

The sample standard deviation provides an estimate of the population standard deviation. The sample standard deviation, s_x, is computed by taking the square root of the sum of the squares of sampled deviations from the mean divided by the sample size minus one.

$$s_x = \sqrt{\frac{1}{n-1}\sum_{i=1}^{n}\left(x_i - \bar{x}\right)^2} \qquad (3\text{-}2)$$

The value *n*-1 is the degrees of freedom for the estimate, which signifies the number of independent pieces of information that go into computing the estimate. Absent any systematic influences during sample collection, the sample standard deviation will approach its population counterpart as the sample size or degrees of freedom increases. The degrees of freedom for an uncertainty estimate is useful for establishing confidence limits and other decision variables.

The sample standard deviation provides an estimate of the repeatability or random error population standard deviation, $\sigma_{\varepsilon_{x,ran}}$. As discussed in Chapter 2, Section 2.4, the standard deviation of an error distribution is equal to the square root of the distribution variance.

$$\sigma_{\varepsilon_{x,ran}} = \sqrt{\text{var}(\varepsilon_{x,ran})} \qquad (3\text{-}3)$$

It has also been shown that

$$u_{\varepsilon_{x,ran}} = \sqrt{\text{var}(\varepsilon_{x,ran})}. \qquad (3\text{-}4)$$

Therefore, the sample standard deviation provides an estimate of the uncertainty due to repeatability or random error.[23]

$$u_{\varepsilon_{x,ran}} \cong s_x. \qquad (3\text{-}5)$$

If the objective of the uncertainty analysis is to characterize a given single measurement

---

[23] The uncertainty due to repeatability or random error in measurement is estimated from a sample of measurements taken over a time period short enough to eliminate variations due to systematic drift or other factors.

performed under specific circumstances, as in developing a statement of capability, then equation (3-5) should be used.

If the estimate is intended to represent the **uncertainty in the mean value** due to repeatability or random error, then the variance of the sample mean is evaluated.

$$\mathrm{var}(\varepsilon_{\bar{x},ran}) = \mathrm{var}(\bar{x})$$

$$= \mathrm{var}\left(\frac{1}{n}\sum_{i=1}^{n} x_i\right) = \frac{1}{n^2}\,\mathrm{var}\left(\sum_{i=1}^{n} x_i\right) \tag{3-6}$$

An important criterion for random sampling is that each of the sampled values must be statistically independent of one another. The variance of a sum of independent variables is the sum of the variances. Therefore, equation (3-6) becomes

$$\mathrm{var}(\varepsilon_{\bar{x},ran}) = \frac{1}{n^2}\sum_{i=1}^{n} \mathrm{var}(x_i) \tag{3-7}$$

Since each x_i is sampled from a population with a variance equal to σ_x^2, then $\mathrm{var}(x_i) = \sigma_x^2$ and

$$\mathrm{var}(\varepsilon_{\bar{x},ran}) = \frac{1}{n^2}\sum_{i=1}^{n} \sigma_x^2 = \frac{\sigma_x^2}{n} \tag{3-8}$$

It has been shown that the population standard deviation σ_x is estimated with the sample standard deviation s_x. Therefore, the uncertainty in the mean value can be estimated to be

$$u_{\varepsilon_{\bar{x},ran}} = \sqrt{\mathrm{var}(\varepsilon_{\bar{x},ran})} \cong \frac{s_x}{\sqrt{n}} \tag{3-9}$$

Once estimates of the sample mean and standard deviation have been obtained, and the degrees of freedom have been noted, it becomes possible to compute limits that bound the sample mean with some specified level of confidence. These limits are called confidence limits and the degree of confidence is called the confidence level.

Confidence limits can be expressed as multiples of the sample standard deviation. For normally distributed samples, this multiple is called the t-statistic. The value of the t-statistic is determined by the desired percent confidence level, C, and the degrees of freedom, ν, for the sample standard deviation.

Confidence limits for a single measured value, x, are given by

$$\bar{x} \pm t_{\alpha/2,\nu} s_x \tag{3-10}$$

and confidence limits for the mean or average, $\bar{x}$, of a sample of measured values are given by

27

$$\overline{x} \pm t_{\alpha/2,v} \frac{s_x}{\sqrt{n}} \qquad (3\text{-}11)$$

where $\alpha = (1 - C/100)$ and $v = n - 1$.

Comparison of equations (3-10) and (3-11) shows that confidence limits about a sample mean value are much smaller than for a single measured value. This should be expected since, with a sample mean, we have more information and a greater expectation of the sample mean value being closer to the population mean.

3.2   Type B Estimates

In some cases, we must attempt to quantify the statistics of measurement error distributions by drawing on our recollected experience concerning the values of measured quantities or on our knowledge of the errors in these quantities.[24] Estimates made in this manner are called heuristic or Type B estimates.

Uncertainty estimates for measurement process errors resulting from reference attribute bias, display resolution, operator bias, computation and environmental factors are typically determined heuristically via containment limits and containment probabilities.

As discussed in Chapter 2, measurement errors can be described by a variety of probability distributions. Of these, the normal and lognormal distributions provide the most realistic statistical representation of measurement errors. Therefore, it is prudent to detail the development of uncertainty estimates for these distributions. Uncertainty estimates for other distributions are discussed in Section 3.2.5.

Computing the degrees of freedom for Type B estimates is discussed in Section 3.2.3. Applying the Student's t distribution for estimating uncertainties with finite degrees of freedom is discussed in Section 3.2.4.

3.2.1   Normal Distribution

If the measurement error is normally distributed, then the uncertainty is computed from

$$u = \frac{L}{\Phi^{-1}\left(\dfrac{1+p}{2}\right)} \qquad (3\text{-}12)$$

where $\pm L$ are the containment limits, p is the containment probability, and $\Phi^{-1}()$ is the inverse normal distribution function.[25]

Containment limits may be taken from manufacturer tolerance limits, stated expanded uncertainties obtained from calibration records or certificates, or statistical process control limits. Containment probability can be obtained from service history data, for example, as the number

---

[24] Information or experience obtained from previous measurement data, general knowledge about the behavior, properties or characteristics of materials or instruments, manufacturer specifications, certificates or other calibration history data, reference data from handbooks, etc.

[25] The inverse normal distribution function can be found in statistics texts and in most spreadsheet programs.

of observed in-tolerances, $n_{\text{in-tol}}$, divided by the number of calibrations, N.

$$C\% = 100\% \frac{n_{\text{in-tol}}}{N}$$

3.2.2   Lognormal Distribution

The lognormal distribution is often used to estimate uncertainty when the error containment limits are asymmetric. The uncertainty is computed from

$$u = |m + q| e^{\lambda^2/2} \sqrt{e^{\lambda^2} - 1} \qquad (3\text{-}13)$$

where q is a physical limit for error the distribution, m is the population median and λ is the shape parameter. The quantities m, q and λ are obtained by numerical iteration, given containment limits and an associated containment probability.

3.2.3   Type B Degrees of Freedom

In equation (3-9), the degrees of freedom are assumed to be infinite. However, we *know* that heuristic estimates are not based on an "infinite" amount of knowledge. As with Type A uncertainty estimates, the degrees of freedom quantifies the amount of information that goes into the Type B uncertainty estimate and is useful for establishing confidence limits and other decision variables.

Therefore, if there is an uncertainty in the containment limits (e.g., $\pm L \pm \Delta L$) or the containment probability (e.g., $\pm p \pm \Delta p$), then it becomes imperative to estimate the degrees of freedom.

Annex G of the GUM provides a relationship for computing the degrees of freedom for a Type B uncertainty estimate

$$\nu \approx \frac{1}{2} \frac{u^2(x)}{\sigma^2 [u(x)]} \approx \frac{1}{2} \left[\frac{\Delta u(x)}{u(x)} \right]^{-2} \qquad (3\text{-}14)$$

where $\sigma^2[u(x)]$ is the variance in the uncertainty estimate, $u(x)$, and $\Delta u(x)$ is the uncertainty in the uncertainty estimate.[26] Hence, the degrees of freedom for a Type B estimate is inversely proportional to the square of the ratio of the uncertainty in the uncertainty divided by the uncertainty.

While this approach is intuitively appealing, the GUM offers no advice about how to determine $\sigma^2[u(x)]$ or $\Delta u(x)$. Since the publication of the GUM, a methodology for determining $\sigma^2[u(x)]$ and computing the degrees of freedom for Type B estimates has been developed. [27] This methodology is outlined in Appendix D.

---

[26] This equation assumes that the underlying error distribution is normal.

[27] Castrup, H.: "Estimating Category B Degrees of Freedom," presented at the 2000 Measurement Science Conference, January 21, 2001.

3.2.4 Student's t Distribution

Once the containment limits, containment probability and the degrees of freedom have been established, we can estimate the standard deviation or uncertainty of the distribution of interest. To do this, we use the Student's t distribution and construct a t-statistic based on the containment probability and degrees of freedom.

The uncertainty estimate is then obtained by dividing the containment limit by the t-statistic, according to equation (3-15).

$$u = \frac{L}{t_{\alpha/2,\nu}} \tag{3-15}$$

3.2.5 Other Distributions

Although the normal, lognormal and Student's t distributions are most often used to estimate uncertainty, other distributions also have limited applicability. As discussed in Chapter 2, many of these distributions are described by minimum bounding limits, $\pm a$ and 100% containment probability (i.e., $p = 1$).

Uncertainty equations for selected distributions are summarized in Table 3-1. Equations for additional distributions are provided in Appendix B.

Table 3-1. Uncertainty Equations for Selected Distributions

Distribution	Distribution Plot	Uncertainty Equation
Quadratic		$u_\varepsilon = \dfrac{a}{\sqrt{5}}$ where $\pm a$ are the minimum bounding limits.
Cosine		$u_\varepsilon = \dfrac{a}{\sqrt{3}}\sqrt{1 - \dfrac{6}{\pi^2}}$ where $\pm a$ are the minimum bounding limits.
Uniform (Rectangular)		$u_\varepsilon = \dfrac{a}{\sqrt{3}}$ where $\pm a$ are the minimum bounding limits.

Distribution	Distribution Plot	Uncertainty Equation
Triangular		$$u_{\varepsilon} = \frac{a}{\sqrt{6}}$$ where $\pm a$ are the minimum bounding limits.
U-Shaped		$$u_{\varepsilon} = \frac{a}{\sqrt{2}}$$ where $\pm a$ are the minimum bounding limits.

CHAPTER 4: INTERPRETING AND APPLYING EQUIPMENT SPECIFICATIONS

Manufacturer specifications are an important element of cost and quality control for testing, calibration and other measurement processes. They are used for equipment selection or establishing equipment substitutions for a given measurement application. In addition, manufacturer specified tolerances are used to compute test uncertainty ratios and estimate bias uncertainties.

Measuring and test equipment (MTE) are periodically calibrated to determine if they are performing within manufacturer specified tolerance limits. In fact, the elapsed-time or interval between calibrations is often based on in-tolerance or out-of-tolerance data acquired from periodic calibrations. Therefore, it is important that manufacturer specifications are properly interpreted and applied.

This chapter discusses how manufacturer specifications are obtained, interpreted and used to assess instrument performance and reliability. Recommended practices and illustrative examples are given for the application to uncertainty estimation. An in-depth discussion about developing, verifying and reporting MTE specifications can be found in NASA *Measurement Quality Assurance* Handbook - Annex 2 *Measuring and Test Equipment Specifications*.

4.1   Measuring and Test Equipment

Before we delve into defining and interpreting specifications, it is important to clarify what constitutes MTE. For the purposes of uncertainty analysis, MTE include artifacts, instruments, sensors and transducers, signal conditioners, data acquisition units, data processors and output displays.

4.1.1   Artifacts

Artifacts constitute passive devices such as mass standards, standard resistors, pure and certified reference materials, gage blocks, etc. Accordingly, artifacts have stated outputs or nominal values and associated specifications.

4.1.2   Instruments

Instruments constitute equipment or devices that are used to measure and/or provide a specified output. They include, but are not limited to, oscilloscopes, wave and spectrum analyzers, Josephson junctions, frequency counters, multimeters, signal generators, simulators and calibrators, inclinometers, graduated cylinders and pipettes, spectrometers and chromatographs, micrometers and calipers, coordinate measuring machines, balances and scales. Accordingly, instruments can consist of various components and associated specifications.

4.1.3   Sensors and Transducers

Sensors constitute equipment or devices that respond to a physical input (i.e., pressure, acceleration, temperature or sound). The terms sensor and transducer are often used interchangeably. Transducers more generally refer to devices that convert one form of energy to another. Consequently, actuators that convert an electrical signal to a physical output are also considered to be transducers. For the purposes of this document, discussion will be limited to sensors and transducers that convert a physical input to an electrical output.

Note: Transmitters constitute sensors coupled with internal signal conditioning and/or data processing components, as well as an output display.

Some sensors and transducers convert the physical input directly to an electrical output, while others require an external excitation voltage or current. Sensors and transducers encompass a wide array of operating principles (i.e., optical, chemical, electrical) and materials of construction. Consequently, their characteristics and associated specifications can cover a broad spectrum of detail and complexity. A selected list of sensors and transducers is shown in Table 4-1.

Table 4-1. Sensors and Transducers

Input	Sensor/Transducer	Output	Excitation
Temperature	Thermocouple RTD Thermistor	Voltage Resistance Resistance	 Current Current, Voltage
Pressure and Sound	Strain Gauge Piezoelectric	Resistance Voltage	Voltage
Force and Torque	Strain Gauge Piezoelectric	Voltage Voltage	Voltage
Acceleration/Vibration	Strain Gage Piezoelectric Variable Capacitance	Voltage Charge Voltage	Voltage Voltage
Position/Displacement	LVDT and RVDT Potentiometer	AC Voltage Voltage	Voltage Voltage
Light Intensity	Photodiode	Current	
Flow Rate	Coriolis Vortex Shedding Turbine	Frequency Pulse/Frequency Pulse/Frequency	 Voltage
pH	Electrode	Voltage	

4.1.4 Signal Conditioners

Signal conditioners constitute devices or equipment that are employed to modify the characteristic of a signal. Conditioning equipment include attenuators, amplifiers, bridge circuits, filters, analog-to-digital and digital-to-analog converters, excitation voltage or current, reference temperature junctions, voltage to frequency and frequency to voltage converters, multiplexers and linearizers. A representative list of signal conditioning methods and functions is provided in Table 4-2.

Table 4-2. Signal Conditioning Methods

Type	Function
Analog-to-Digital Conversion (ADC)	Quantization of continuous signal
Amplification	Increase signal level

Type	Function
Attenuation	Decrease signal level
Bridge Circuit	Increase resistance output.
Charge Amplification	Convert charge to voltage.
Cold Junction Compensation	Provide temperature correction for thermocouple connection points.
Digital-to-Analog Conversion (DAC)	Convert discrete signal to continuous signal
Excitation	Provide voltage or current to transducer.
Filter	Provide frequency cutoffs and noise reduction
Isolation	Block high voltage and current surges.
Linearization	Convert non-linear signal to representative linear output.
Multiplexing	Provide sequential routing of multiple signals.

4.1.5 Data Acquisition

Data acquisition (DAQ) equipment provide the interface between the signal and the data processor or computer. DAQ equipment include high speed timers, random access memory (RAM) and cards containing signal conditioning components.

4.1.6 Data Processors

Data processors constitute equipment or methods used to implement necessary calculations. Data processors include totalizers and counters, statistical methods, regression or curve fitting algorithms, interpolation schemes, measurement unit conversion or other computations. Error sources resulting from data reduction and analysis are often overlooked in the assessment of measurement uncertainty.

4.1.7 Output Displays

Output display devices constitute equipment used to visually present processed data. Display devices can be analog or digital in nature. Analog devices include chart recorders, plotters and printers, dials and gages, cathode ray tube (CRT) panels and screens. Digital devices include light-emitting diode (LED) and liquid crystal display (LCD) panels and screens. Resolution is a primary source of error for digital and analog displays.

4.2 Performance Characteristics

Manufacturer specifications should provide an objective assessment of MTE performance characteristics. However, understanding specifications and using them to compare or select equipment from different manufacturers or vendors can be a difficult task. This primarily results from inconsistent terminology, units, and methods used to develop and report equipment specifications.

Some manufacturers may provide ample information detailing individual performance specifications, while others may only provide a single specification for overall accuracy. In some instances, specifications can be complex, including numerous time or range dependent characteristics. And, since specification documents are also a means for manufacturers to market their products, they often contain additional information about features, operating condition limits, or other qualifiers.

4.2.1 Static Characteristics

Static performance characteristics provide an indication of how an instrument, transducer or signal conditioning device responds to a steady-state input at one particular time. In addition to sensitivity (or gain) and zero offset, other static characteristics include nonlinearity, repeatability, hysteresis, resolution, noise, transverse sensitivity, acceleration sensitivity, thermal stability, thermal sensitivity shift, temperature drift, thermal zero shift, temperature coefficient, and accuracy.[28]

4.2.2 Dynamic Characteristics

Dynamic performance characteristics provide an indication of how an instrument, transducer or signal conditioning device responds to changes in input over time. Dynamic characteristics include warm-up time, response time, time constant, settling time, zero drift, sensitivity drift, stability, upper and lower cutoff frequencies, bandwidth, resonant frequency, frequency response, damping, phase shift, and reliability.[29]

4.2.3 Other Characteristics

Other characteristics are often included with performance specifications to indicate input and output ranges, environmental operating conditions, external power requirements, weight, dimensions and other physical aspects of the device. These other characteristics include rated output, full scale output, range, span, dynamic input range, threshold, dead band, operating temperature range, operating pressure range, operating humidity range, storage temperature range, thermal compensation, temperature compensation range, vibration sensitivity, excitation voltage or current, weight, length, height, and width.

4.3 Obtaining Specifications

Manufacturers publish MTE specifications on their web pages, in product data sheets, technical notes, control drawings and operating manuals. Some manufacturers also maintain an archive of specification information for discontinued products. In some instances, manufacturers will only provide MTE specification information upon formal request by phone, fax or email. In general, however, published specifications are relatively easy to find via an internet search.

4.4 Interpreting Specifications

Ultimately, the MTE user must determine which specifications are relevant to their application. Therefore, a basic understanding of the fundamental operating principles of the MTE is an important requirement for proper interpretation of performance specifications. In some cases, first-hand experience about the MTE may be gained through calibration and testing. In other cases, detailed knowledge about the MTE may be obtained from operating manuals, training courses, patents and other technical documents provided by the manufacturer.

Ideally, MTE specifications provide adequate details about the expected performance characteristics of a representative group of identical devices or items (i.e., a specific manufacturer and model). This information should be reported in a logical format, using

---

[28] Accuracy is typically reported as a combined specification that accounts for nonlinearity, hysteresis, and repeatability.

[29] Reliability specifications typically refer to performance over an extended time-period or maximum number of cycles.

consistent terms, abbreviations and units that clearly convey pertinent performance characteristics.

For the most part, manufacturer specifications are intended to convey tolerances or limits that are expected to bound the MTE performance characteristics. For example, these limits may correspond to temperature, shock and vibration parameters that affect the sensitivity and/or zero offset of a sensing device.

Unfortunately, there is no universal guide or standard regarding the development and reporting of MTE specifications. Inconsistency in the methods used to develop and report performance specifications, and in the terms and units used to convey this information, create obstacles to the proper understanding and interpretation of MTE specifications.

In select instances, the information included in a specification document may follow a standardized format.[30] However, the vast majority of specification documents fall short of providing crucial information about the confidence levels associated with reported specification limits. MTE manufacturers also don't indicate the applicable probability distribution for a particular performance characteristic.

Consequently, it is difficult to estimate uncertainties from MTE specifications without gaining further clarification or making some underlying assumptions. It is a good practice to

1. Review the specifications and highlight the MTE characteristics that need clarification.
2. Check the operating manual and associated technical documents for other useful details.
3. Request additional information and clarification from the manufacturer's technical department.

4.4.1 Terms, Definitions and Abbreviations

Technical organizations, such as ISA and SMA, have published documents that adopt standardized instrumentation terms and definitions.[31,32] However, there is a need for further clarification and consistency in the general terms and definitions used in the reporting of MTE specifications. General terms and definitions for MTE specifications and other related characteristics are provided in Appendix A. There are particular terms and abbreviations that require further discussion.

For example, some MTE specifications may convey performance characteristics as "typical" or "maximum" values. However, the basis for these classifications is not often apparent and introduces confusion about which specification (typical or maximum) is applicable. In addition, since associated confidence levels, containment probabilities or coverage factors are not often

---

[30] See for example, ISA-RP37.2-1982-(R1995): *Specifications and Tests for Strain Gauge Pressure Transducers*, The Instrumentation, Systems and Automation Society, Reaffirmed December 14, 1982.

[31] ISA-37.1-1975 (R1982): *Electrical Transducer Nomenclature and Terminology*, The Instrumentation, Systems and Automation Society, Reaffirmed December 14, 1982.

[32] SMA LCS 04-99: *Standard Load Cell Specifications*, Scale Manufacturers Association, Provisional 1st Edition, April 24, 1999.

provided, it is difficult to clearly interpret either set of specifications. Consequently, the manufacturer must be contacted for further clarification.

MTE specifications commonly include the use of abbreviations such as FS, FSO, FSI, RDG, RO, RC and BSL. The abbreviation FS (or F.S.) refers to full scale. Similarly, the abbreviation FSO (or F.S.O.) refers to full scale output and the abbreviation FSI (or F.S.I.) refers to full scale input. Specifications that are reported as % FS (or ppm FS) generally refer to full scale output. When in doubt, however, contact the manufacturer for clarification.

The abbreviation RDG refers to reading or output value. The abbreviation RO (or R.O.) refers to rated output and the abbreviation RC (or R.C.) refers to rated capacity. Some MTE specifications also use the abbreviation BSL (or B.S.L.) to indicate that a combined non-linearity, hysteresis, and repeatability specification is based on observed deviations from a best-fit straight line. Abbreviations commonly used in MTE specifications are listed in the Acronyms and Abbreviations section of this document.

4.4.2 Qualifications, Stipulations and Warnings

Most MTE specifications describe the performance characteristics covered by the manufacturer's product warranty. These reported specifications also often include qualifications, clarifications and/or caveats. Therefore, it is a good practice to read all notes and footnotes carefully to determine which, if any, are relevant to the specifications.

For example, MTE specification documents commonly include a footnote warning that the values are subject to change or modification without notice. Manufacturers do not generally modify existing MTE specifications unless significant changes in components or materials of construction warrant the establishment of new specifications. However, it may be necessary to contact the manufacturer to ensure that the appropriate MTE specification documents are obtained and applied.[33]

MTE specifications may state a recommended range of environmental operating conditions to ensure proper performance. They may also include a qualification indicating that all listed specifications are typical values referenced to standard conditions (e.g., 25 °C and 10 VDC excitation). This qualification implies that the primary performance specifications were developed from tests conducted under a particular set of conditions.

If so, additional specifications, such as thermal zero shift, thermal sensitivity shift and thermal transient response error, are included to account for the variation in actual MTE operating conditions from standard conditions. The MTE user must then consider whether or not these additional specifications are relevant to the MTE application.

4.4.3 Specification Units

As with terms and definitions, specification units can vary between manufacturers of similar MTE models. In addition, specification units can vary from one performance characteristic to another for a given MTE manufacturer model.

---

[33] That is, the published specifications considered by the manufacturer to be applicable at the time the MTE was purchased.

For example, display resolution specifications can be expressed in digits, counts, percent (%) or other units such as mV or °C. Nonlinearity, hysteresis and repeatability specifications can be expressed as % FS, ppm FS, % RDG, ppm RDG, % RO or other units. Sensitivity specifications can be expressed as mV/psi,
an be expressed as % FS/°F, % RO/°C, ppm/°C, % FS/g, psi/g, psi/°F, mV/°C, %Load/°F, etc. Noise specifications such as Normal Mode Rejection Ratio (NMRR) and Common Mode Rejection Ratio (CMRR) are generally specified in decibels (dB) at specified frequencies (usually 50 and 60 Hz).

Different specification units can make it especially difficult to interpret specifications. In most cases, unit conversion is required before specifications can be properly applied. Selected specification conversion factors are listed in Table 4-3 for illustration.

Table 4-3. Specification Conversion Factors

Percent	ppm	dB	Relative to 10 V	Relative to 100 psi	Relative to 10 kg/°C
1%	10000	-40	100 mV	1 psi	100 g/°C
0.1%	1000	-60	10 mV	0.1 psi	10 g/°C
0.01%	100	-80	1 mV	0.01 psi	1 g/°C
0.001%	10	-100	100 μV	0.001 psi	100 mg/°C
0.0001%	1	-120	10 μV	0.0001 psi	10 mg/°C

Note: A decibel (dB) is a dimensionless unit for expressing the ratio of two values of power, P_1 and P_2, where $dB = 10 \log(P_2/P_1)$. The dB values in Table 4-3 are computed for P_2/P_1 ratios corresponding to the percent and ppm values listed. For electrical power, it is important to note that power is proportional to the square of voltage, V, so that $dB_V = 10 \log (V_1^2/V_2^2) = 20 \log (V_1/V_2)$. Similarly, acoustical power is proportional to the square of sound pressure, p, so that $dB_A = 10 \log (p_1^2/p_2^2) = 20 \log (p_1/p_2)$.

Additional calculations may be required before specifications can be properly used to estimate MTE parameter bias uncertainty and tolerance limits. This brings us to the topic of applying specifications.

4.5   Applying Specifications

Manufacturer specifications can be used to purchase or substitute MTE for a given measurement application, estimate bias uncertainties and establish tolerance limits for calibration and testing. Therefore, MTE users must be proficient at identifying applicable specifications and in interpreting and combining them.

It is also important that manufacturers and users have a good understanding and assessment of the confidence levels and error distributions applicable to MTE specifications. This is a crucial part of the process and requires some further discussion.

4.5.1   Confidence Levels

Some manufacturer MTE specifications are established by testing a sample of the produced model population. The sample test results are used to develop limits that ensure a large percentage of the MTE model population will perform as specified. Consequently, the specifications are confidence limits with associated confidence levels.[34]

That is, the limits specified for an MTE performance characteristic are established for a particular confidence level and degrees of freedom (or sample size), as discussed in Chapter 2. Confidence limits, $\pm L_x$, for values of a specific performance characteristic, x, are expressed as

$$\pm L_x = \pm t_{\alpha/2,v} s_x \qquad \qquad (4\text{-}1)$$

where

$$
\begin{aligned}
t_{\alpha/2,v} &= \text{t-statistic} \\
\alpha &= \text{significance level} = 1 - C/100 \\
C &= \text{confidence level (\%)} \\
v &= \text{degrees of freedom} = n - 1 \\
n &= \text{sample size} \\
s_x &= \text{sample standard deviation.}
\end{aligned}
$$

Ideally, confidence levels should be commensurate with what MTE manufacturers consider to be the maximum allowable false accept risk (FAR).[35] The general requirement is to minimize the probability of shipping an MTE item with nonconforming (or out-of-compliance) performance characteristics. In this regard, the primary factor in setting the maximum allowable FAR may be the costs associated with shipping nonconforming products.

Unfortunately, manufacturers don't commonly report confidence levels for their MTE specifications. In fact, the criteria and motives used by manufacturers to establish MTE specifications are not often apparent. Most MTE manufacturers see the benefits, to themselves and their customers, of establishing specifications with high confidence levels. However, competition between MTE manufacturers can result in unrealistically optimistic specifications that, in-turn, can result in excessive out-of-tolerance occurrences.[36]

Alternatively, some manufacturers may test the entire produced MTE model population to ensure that individual items are performing within specified limits prior to shipment. However, this compliance testing process does not ensure a 100% probability (or confidence level) that the customer will receive an in-tolerance item. The reasons for this include

1. Measurement uncertainty associated with the manufacturer MTE compliance testing process.
2. MTE bias drift or shift resulting from shock, vibration and other environmental extremes during shipping and handling.

Manufacturers may account for the uncertainty in their testing and measurement processes by using a higher confidence level (e.g., 99.9%) to establish larger specification limits or by

---

[34] In this context, confidence level and containment probability are synonymous, as are confidence limits and containment limits.
[35] From a producer or manufacturer's perspective, false accept risk is the probability of accepting and shipping a nonconforming item.
[36] See for example, Deaver, David: "Having Confidence in Specifications," proceeding of NCSLI Workshop and Symposium, Salt Lake City, UT, July 2004.

employing arbitrary guardbanding[37] methods and multiplying factors. In either case, the resulting MTE specifications are not equivalent to 100% confidence limits.

Some manufacturers also conduct special environmental and accelerated life testing on a population subset to quantify the effects of potential shipping and handling stresses. They might even include separate specifications for these effects. However, not all MTE manufacturers incorporate these rigorous practices.

4.5.2   Error Distributions

MTE performance characteristics, such as nonlinearity, repeatability, hysteresis, resolution, noise, thermal stability and zero shift constitute sources of measurement error. As discussed in Chapter 2, measurement errors are random variables that follow probability distributions. Therefore, MTE performance characteristics are also considered to be random variables that follow probability distributions.

This concept is important to the interpretation and application of MTE specifications because an error distribution allows us to determine the probability that a performance characteristic is in conformance with its specification.

Typically, manufacturers do not identify an underlying distribution for performance specifications. This might imply that a specification simply bounds the range of values. For the sampled MTE model specifications described in section 4.5.1., the performance characteristics of an individual unit may vary from the population mean. However, the majority of the units should perform well within the specification limits. Accordingly, a central tendency exists that can be described by the normal distribution.

If the limits are asymmetric about a specified nominal value, it is still reasonable to assume that individual MTE performance characteristics will tend to be distributed near the nominal value. In this case, the normal distribution may still apply. However, the lognormal or other asymmetric distribution may be more applicable.

There are a couple of exceptions when the uniform distribution would be applicable. These include digital output resolution error and quantization error resulting from the digital conversion of an analog signal. In these instances, the specifications limits, , $\pm L_{res}$ and $\pm L_{quan}$, would be 100% confidence limits defined as

$$\pm L_{res} = \pm \frac{h}{2} \tag{4-2}$$

and

$$\pm L_{quan} = \pm \frac{A}{2^{n+1}} \tag{4-3}$$

where

h = least significant display digit
A = full scale range of analog to digital converter
n = quantization significant bits.

---

[37] Guardbands are supplemental limits used to reduce false accept risk during calibration and testing.

4.5.3 Combining Specifications

In testing and calibration processes, an MTE performance characteristic is identified as being in-tolerance or out-of-tolerance. In some cases, the tolerance limits are determined from a combination of MTE specifications. For example, consider the accuracy specifications for the DC voltage function of a Fluke 8062A digital multimeter.[38]

For a displayed reading of 5 VDC, the accuracy specification is reported as $\pm$ (0.07% Reading + 2 digits) and the resolution as 1 mV. In this case, the accuracy specification is $\pm$ (0.07% Reading + 2 mV).[39] To compute the combined accuracy specification, we must convert the % Reading to a value in mV units.

$$0.07\% \text{ Reading} = (0.07/100) \times 5 \text{ V} \times 1000 \text{ mV/V} = 3.5 \text{ mV}$$

The total accuracy specification for the 5 V output reading would then be $\pm$ (2.5 mV + 2 mV) or $\pm$ 5.5 mV.

For another example, consider the tolerance specifications for different gage block grades published by NIST.[40] Suppose we want to compute the combined tolerance limits for a Grade 2 gage block with 20 mm nominal length. There are two sets of specification limits. The first specification limits (+0.10 μm, -0.05 μm) are asymmetric, while the second specification limits ($\pm$ 0.08 μm) are symmetric. Consequently, the combined tolerance limits will be asymmetric and upper and lower tolerances (e.g., $+L_1$, $-L_2$) must be computed.

There are two possible ways to compute values for L_1 and L_2 from the specifications: linear (additive) combination or root sum square (RSS) combination.

1. Linear Combination

$$L_1 = 0.10 + 0.08 = 0.18$$
$$L_2 = 0.05 + 0.08 = 0.13$$

2. RSS Combination

$$L_1 = \sqrt{(0.10)^2 + (0.08)^2} = \sqrt{0.0164} = 0.13$$
$$L_2 = \sqrt{(0.05)^2 + (0.08)^2} = \sqrt{0.0089} = 0.09$$

If the specifications are interpreted to be additive, then the combined tolerance limits for the 20 mm Grade 2 gage block are +0.18 μm, -0.13 μm. Alternatively, if they are combined in RSS, then the resulting tolerance limits are +0.13 μm, -0.09 μm.

---

[38] Specifications from 8062A Instruction Manual downloaded from www.fluke.com
[39] Understanding Specifications for Precision Multimeters, Application Note Pub_ID 11066-eng Rev 01, ©2006 Fluke Corporation.
[40] The Gage Block Handbook, NIST Monograph 180, 1995.

Linear or RSS specification combination cannot be used for MTE that have complex performance characteristics. For example, consider the specifications for a Transducer Techniques MDB-5-T load cell .[41]

The load cell sensing element is a resistance-based strain gauge that requires an external excitation voltage. This load cell has a rated output of 2 mV/V for loads up to 5 lb$_f$ which equates to a nominal sensitivity of 0.4 mV/V/lb$_f$. Therefore, the load cell output is a function of the excitation voltage and the applied load.

$$LC_{out} = W \times S \times V_{Ex} \tag{4-4}$$

where

W = Applied load or weight
S = Load cell sensitivity
V_{Ex} = Excitation voltage

Equation (4-4) shows the mathematical relationship between the physical input (i.e., weight) and the electrical output (i.e., voltage) of the load cell.[42] This relationship is called a transfer function.

According to the specifications, the load cell output will be affected by the following error sources:

- Excitation Voltage, ± 0.25 V
- Nonlinearity, ± 0.05% of R.O.
- Hysteresis, ± 0.05% of R.O.
- Noise, ± 0.05% of R.O.
- Zero Balance, ± 1% of R.O.
- Temperature Effect on Output, ± 0.005% of Load/°F
- Temperature Effect on Zero, ± 0.005% of R.O./°F

If the load cell is tested or calibrated using a weight standard, then any error associated with the weight should also be included.

Equation (4-4) needs to be modified to account for these error sources. Unfortunately, given the assortment of specification units, the error terms cannot simply be added at the end of the equation. The appropriate load cell output equation is expressed in equation (4-5).

$$LC_{out} = \left[\left(W_s + TE_{out} \times TR_{\circ F} \right) \times S + NL + Hys + NS + ZO + TE_{zero} \times TR_{\circ F} \right] \times V_{Ex} \tag{4-5}$$

where

$$W_s = W_n + W_e \tag{4-6}$$

---

[41] Specifications obtained from www.ttloadcells.com/mdb-load-cell.cfm
[42] The validity of this equation depends on the use of appropriate units for the variables, W, S and V_{Ex}.

$$V_{Ex} = V_n + V_e \qquad\qquad (4\text{-}7)$$

and

W_n = Nominal or stated value of weight standard
W_e = Bias of weight standard
V_n = Nominal excitation voltage
V_e = Excitation voltage error
TE_{out} = Temperature effect on output
$TR_{\circ F}$ = Temperature range in °F
NL = Nonlinearity
Hys = Hysteresis
NS = Noise and ripple
ZO = Zero offset
TE_{zero} = Temperature effect on zero

Equations (4-5) through (4-7) constitute an error model for the load cell output. As discussed in Chapter 2, given some knowledge about the error distributions, the variance addition rule can be applied to estimate the uncertainty in the load cell output voltage for a given applied load.

This procedure involves some additional concepts and methods that are covered in subsequent chapters. A detailed uncertainty analysis of a load measurement system is presented in Chapter 7.

CHAPTER 5: DIRECT MEASUREMENTS

In direct measurements, the quantity of interest (i.e., subject parameter or measurand) is obtained directly by measurement and is not determined indirectly by computing its value from the measurement of other variables or quantities. Examples of direct measurements include, but are not limited to the following:

- Measuring the length of an object with a ruler or micrometer.

- Measuring the output from a DC voltage reference with a voltmeter.

- Measuring the temperature of a substance using a liquid-in-glass thermometer.

In this chapter, the analysis of a micrometer calibrated with a gage block is used to illustrate the basic concepts and methods used to estimate uncertainty for direct measurements. The general uncertainty analysis procedure includes the following the steps outlined in Chapter 2:

1. Define the Measurement Process
2. Develop the Error Model
3. Identify Error Sources and Distributions
4. Estimate Uncertainties
5. Combine Uncertainties
6. Report Analysis Results

5.1   Define the Measurement Process

In this example, a 0-25 mm digital micrometer is calibrated at 10 mm nominal length using a Class 2 (Grade 2) gage block set. Multiple readings of the 10 mm gage block length are taken with the micrometer. The repeat readings observed with the micrometer are listed in Table 5-1.

Table 5-1. Micrometer Measurements

Reading	Length (mm)	Deviation from Nominal (μm)
1	10.003	3
2	10.002	2
3	10.003	3
4	10.004	4
5	10.001	1
6	10.005	5
7	10.002	2
8	10.004	4

In this analysis, the quantity of interest is the average length obtained from the micrometer measurements corrected to a standard reference temperature of 20 °C. This value will be reported along with its estimated total uncertainty. The results of the uncertainty analysis will be used to determine if the micrometer is within the manufacturer specified tolerance limits.

5.1.1 Gage Block Specifications

The tolerance specifications for the Grade 2 gage block set are obtained from tabulated data published by NIST.[43] Subsets of the data are listed in Tables 5-2 and 5-3.

Table 5-2. Tolerance Grades for Metric Gage Blocks (μm)

Nominal	Grade .5	Grade 1	Grade 2	Grade3
< 10 mm	0.03	0.05	+0.10, -0.05	+0.20, -0.10
< 25 mm	0.03	0.05	+0.10, -0.05	+0.30, -0.15
< 50 mm	0.05	0.10	+0.20, -0.10	+0.40, -0.20
< 75 mm	0.08	0.13	+0.25, -0.13	+0.45, -0.23
< 100 mm	0.10	0.15	+0.30, -0.15	+0.60, -0.30

Table 5-3. Additional Tolerance for Length, Flatness, and Parallelism (μm)

Nominal	Grade .5	Grade 1	Grade 2	Grade3
< 100 mm	± 0.03	± 0.05	± 0.08	± 0.10
< 200 mm		± 0.08	± 0.15	± 0.20
< 300 mm		± 0.10	± 0.20	± 0.25
< 500 mm		± 0.13	± 0.25	± 0.30

Gage block length is defined at the following standard reference conditions:

temperature	=	20 °C (68 °F)
barometric pressure	=	101.325 KPa (14.7 psia)
water vapor pressure	=	1.33 KPa (10 mm of mercury)
CO_2 content of air	=	0.03%.

Only temperature has a measurable effect on the physical length of the gage block as a result of thermal expansion or contraction. The nominal coefficient of thermal expansion for gage block steel is 11.5×10^{-6}/°C. According to ANSI/ASME,[44] the maximum allowable limits for the coefficient of thermal expansion are $\pm 1 \times 10^{-6}$/°C.

5.1.2 Micrometer Specifications

Manufacturer specifications for the micrometer state a digital resolution of 1 μm and error (tolerance) limits of ± 4 μm. For the purposes of this analysis, the coefficient of thermal expansion for the micrometer is taken to be 5.6×10^{-6}/deg °C with corresponding error limits of $\pm 0.5 \times 10^{-6}$/°C.

5.1.3 Environmental Temperature Specifications

During the measurement process, an average laboratory temperature of 23 °C was monitored and maintained. The tolerance limits of the temperature monitoring device are ± 2 °C.

---

[43] The Gage Block Handbook, NIST Monograph 180, 1995.

[44] Precision Gage Blocks for Length Measurement (Through 20 in. and 500 mm), ANSI/ASME B89.1.9M-1984.

5.2   Define the Error Model

In this example, a 10 mm nominal gage block is measured with a micrometer and the average length reported. Therefore, the basic measurement model for the length, x, is defined as

$$x = x_{true} + \varepsilon_x \qquad (5\text{-}1)$$

where

$$\begin{aligned}
x_{true} &= \text{ true gage block length} \\
\varepsilon_x &= \text{ total error in the length measurement.}
\end{aligned}$$

The error model for ε_x is the sum of the errors encountered during the length measurement process and can be generally expressed as

$$\varepsilon_x = \varepsilon_1 + \varepsilon_2 + \cdots + \varepsilon_n \qquad (5\text{-}2)$$

where the numbered subscripts signify the different error sources.

5.3   Identify Error Sources and Distributions

In the length measurement process, we must account for the following errors:

- Bias in the value of the 10 mm gage block length, ε_{Gbias}.

- Error associated with repeat measurements, ε_{ran}.

- Error associated with the digital resolution of the micrometer, ε_{Mres}.

- Operator bias during the micrometer measurement process, ε_{op}.

- Environmental factors errors resulting from thermal expansion of the gage block and the micrometer, ε_{env}.

The micrometer bias is not included, because this is what is estimated in the uncertainty analysis. The error model for the length measurement can now be expressed as

$$\varepsilon_x = \varepsilon_{Gbias} + \varepsilon_{ran} + \varepsilon_{Mres} + \varepsilon_{op} + \varepsilon_{env}. \qquad (5\text{-}3)$$

The specifications for the gage block and micrometer do not provide insight about which probability distribution to apply to each of these error sources. However, as discussed in Chapter 4, Section 4.5.2, error distributions often exhibit a central tendency.

In general, if an error distribution has a central tendency and the error limits are symmetric, the normal distribution is applicable. If the error limits are not symmetric, the lognormal or other asymmetric distribution may be more applicable. For the length measurement example, the uniform distribution is only applicable to the micrometer digital resolution error. More discussion on the selection and application of error distributions for the length measurement example are discussed in the following section.

5.4   Estimate Uncertainties

With the exception of repeatability or random error, the uncertainty in each error source must be estimated heuristically from the containment limits, $\pm L$, containment probability, p, and the inverse error distribution function, $F^{-1}(p)$, as shown in equation (5-4).

$$u = \frac{L}{F^{-1}(p)}$$

(5-4)

As discussed in Chapter 4, equipment specifications should convey key information about the performance characteristics of the MTE. For the most part, manufacturer specification data include $\pm$ limits for error sources that affect the MTE performance. Information about the confidence level associated with these specification limits or the applicable error distribution are not often provided.

Consequently, it is a good practice to thoroughly review the appropriate MTE specification information and highlight items that need clarification. The manufacturer should then be contacted for additional information and clarification as required. If this information is not obtainable from the manufacturer, then alternative sources should be employed including your own experience and best judgement.

5.4.1   Gage Block Bias

The gage block specifications indicate that the length bias is comprised of two error sources

$$\varepsilon_{bias} = \varepsilon_{tol} + \varepsilon_{lfp}$$

(5-5)

where ε_{tol} is the tolerance error and ε_{lfp} is the error due to length, flatness and parallelism. Applying the variance addition rule,

$$\begin{aligned} var(\varepsilon_{bias}) &= var(\varepsilon_{tol} + \varepsilon_{lfp}) \\ &= var(\varepsilon_{tol}) + var(\varepsilon_{lfp}) + 2cov(\varepsilon_{tol}, \varepsilon_{lfp}) \end{aligned}$$

(5-6)

where $cov(\varepsilon_{tol}, \varepsilon_{lfp})$ is the covariance between ε_{tol} and ε_{lfp}. From Axiom 2 and equation (5-6), the gage block bias uncertainty can be expressed as

$$u_{Gbias} = \sqrt{u_{tol}^2 + u_{lfp}^2 + 2\rho_{tol,lfp}\, u_{tol} u_{lfp}}\ .$$

(5-7)

The tolerance error limits for Grade 2 gage blocks with nominal length less than 25 mm are $+ 0.10\ \mu m$ and $- 0.05\ \mu m$. Given these skewed limits, the lognormal distribution should be applicable for ε_{tol}. The error limits for length, flatness and parallelism for Grade 2 gage blocks with nominal length less than 100 mm are $\pm 0.08\ \mu m$. Therefore, the normal distribution should be applicable for ε_{lfp}.

From experience, we know that gage block specifications typically represent a high in-tolerance or containment probability. In this analysis, we will assume that a 99% containment probability applies for both error source limits.

5.4.1.1   Tolerance Error

To compute the uncertainty in the tolerance error, u_{tol}, we refer to the lognormal distribution plot shown in Figure 5-1, where the mode, M, is equal to 10 mm, the lower containment limit, L_1, is – 0.05 μm, the upper containment limit, L_2, is + 0.10 μm and the containment probability is 99%.

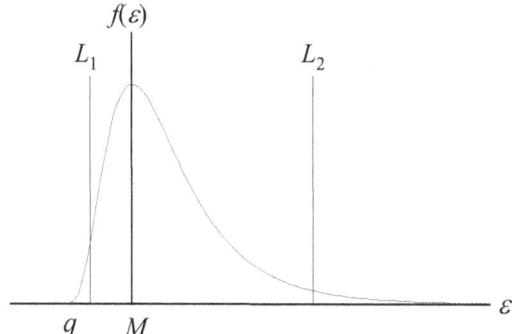

Figure 5-1. Right-handed Lognormal Distribution

As discussed in Appendix B, section B.2, the probability density function for a right-handed lognormal distribution is given by

$$f(\varepsilon) = \frac{1}{\sqrt{2\pi}\,\lambda\,|\varepsilon - q|}\exp\left\{-\left[\ln\left(\frac{\varepsilon - q}{m - q}\right)\right]^2 \Big/ 2\lambda^2\right\}$$

where q is the physical limit for ε, m is the population median and λ is the shape parameter. The uncertainty, u_{tol}, is the population standard deviation, σ, which is defined as

$$\sigma = |m - q|\,e^{\lambda^2/2}\sqrt{e^{\lambda^2} - 1}\,.$$

The population median is defined as

$$m = q\left(1 - e^{\lambda^2}\right).$$

The unknown variables q and λ must be solved for iteratively using the containment limits and containment probability.[45] The numerical iteration was conducted off-line and the resulting uncertainty estimated to be

$$u_{tol} = 0.0287 \text{ μm}.$$

5.4.1.2   Length, Flatness and Parallelism Error

The uncertainty due to gage block length, flatness and parallelism error can be computed from the ± 0.08 μm containment limits, 99% containment probability and the inverse normal distribution function, $\Phi^{-1}(\cdot)$. The inverse normal distribution function, can be found in statistics

---

[45] Additional guidance is provided in Appendix B, Section B.2.

texts and in most spreadsheet programs.

$$u_{lfp} = \frac{0.08 \; \mu m}{\Phi^{-1}\left(\dfrac{1+0.99}{2}\right)} = \frac{0.08 \; \mu m}{2.5758} = 0.0311 \; \mu m.$$

5.4.1.3   Gauge Block Bias Uncertainty

There is no reason to believe that there is any correlation between the gage block tolerance error and the error due to length, flatness and parallelism. Therefore, the total uncertainty in the gage block bias is estimated to be

$$
\begin{aligned}
u_{Gbias} &= \sqrt{(0.0287 \; \mu m)^2 + (0.0311 \; \mu m)^2} \\
&= \sqrt{0.00179 \; \mu m^2} \\
&= 0.042 \; \mu m.
\end{aligned}
$$

5.4.2   Repeatability (Random Error)

The uncertainty in the repeatability or random error in the length measurement is determined from the repeat measurements. As discussed in Chapter 3, the uncertainty due to repeatability is equal to the standard deviation of the sample data. The standard deviation of the sample of length measurements is given by

$$s_x = \sqrt{\frac{1}{n-1} \sum_{i=1}^{n} (x_i - \bar{x})^2}$$

where x_i is the ith reading and the mean value of the sample is computed from

$$\bar{x} = \frac{1}{n}(x_1 + x_2 + \ldots + x_n).$$

The mean value of the length measurements is

$$\bar{x} = \frac{(10.003 + 10.002 + 10.003 + 10.004 + 10.001 + 10.005 + 10.002 + 10.004)}{8} \; mm$$

$$= 10.003 \; mm$$

and the differences between the measured values and the mean value are

$$x_1 - \bar{x} = 10.003 - 10.003 = 0.000 \text{ mm} = 0 \text{ μm}$$

$$x_2 - \bar{x} = 10.002 - 10.003 = -0.001 \text{ mm} = -1 \text{ μm}$$

$$x_3 - \bar{x} = 10.003 - 10.003 = 0.000 \text{ mm} = 0 \text{ μm}$$

$$x_4 - \bar{x} = 10.004 - 10.003 = 0.001 \text{ mm} = 1 \text{ μm}$$

$$x_5 - \bar{x} = 10.001 - 10.003 = -0.002 \text{ mm} = -2 \text{ μm}$$

$$x_6 - \bar{x} = 10.005 - 10.003 = 0.002 \text{ mm} = 2 \text{ μm}$$

$$x_7 - \bar{x} = 10.002 - 10.003 = -0.001 \text{ mm} = -1 \text{ μm}$$

$$x_8 - \bar{x} = 10.004 - 10.003 = 0.001 \text{ mm} = 1 \text{ μm}.$$

The standard deviation is

$$s_x = \sqrt{\frac{(0)^2 + (-1)^2 + (0)^2 + (1)^2 + (-2)^2 + (2)^2 + (-1)^2 + (1)^2}{7}} \text{ μm}$$

$$= \sqrt{\frac{12}{7}} \text{ μm} = \sqrt{1.71} \text{ μm} = 1.31 \text{ μm}.$$

Repeatability uncertainty is

$$u_{x,ran} = s_x = 1.31 \text{ μm}$$

and the repeatability uncertainty in the mean value is

$$u_{\bar{x},ran} = \frac{s_x}{\sqrt{n}} = \frac{1.31 \text{ μm}}{\sqrt{8}} = 0.463 \text{ μm}.$$

Since the mean value is the quantity of interest in this analysis, $u_{\bar{x},ran}$ should be included in the overall uncertainty estimate.

5.4.3 Resolution Error

To estimate the uncertainty due to resolution error, we note that the micrometer has a digital readout. Therefore, the resolution error can be assumed to be uniformly distributed. The resolution uncertainty is estimated from the ± 0.5 μm containment limits, 100% containment probability and the inverse uniform distribution function.

$$u_{res} = \frac{0.5 \text{ μm}}{\sqrt{3}} = 0.289 \text{ μm}$$

5.4.4 Operator Bias

Inconsistencies as the operator uses the micrometer to measure the gage block length are most likely accounted for in the repeatability or random error. However, we still need to account for the possibility of some consistent or systematic operator bias during the measurement process. Some possible sources of operator bias include how the operator positions the micrometer on the gauge block and the amount of clamping force applied to the gage block.

50

Since we do not know the sign or magnitude of the operator bias, we consider it to be a normally distributed error source. To estimate the uncertainty in the operator bias, we will assume containment limits that are based on half of the resolution, with a 90% containment probability.

$$u_{op} = \frac{(0.5)(1\ \mu m)}{\Phi^{-1}\left(\dfrac{1+0.90}{2}\right)} = \frac{0.5\ \mu m}{1.6449} = 0.304\ \mu m\ .$$

> **Note:** The containment limits for the operator bias are not necessarily based on resolution error. Best judgement and knowledge should be used in developing appropriate containment limits and containment probability.

5.4.5   Environmental Factors Error

For this error source, we are interested in determining the uncertainty in the length measurement due to thermal expansion effects. In this case, we must consider the thermal expansion of the gage block and the micrometer. We must also account for the uncertainty in the environmental temperature reading and the uncertainty in the expansion coefficients.

The change in length measurement, Δx, due to the temperature departure from 20 °C nominal, results from the expansion (or contraction) of the gage block and the micrometer. The net change is computed from the following equation

$$\Delta x = x_{nom} \times (\alpha_g - \alpha_m) \times \Delta T \tag{5-8}$$

where

$$
\begin{aligned}
x_{nom} &= \text{nominal gage block length} = 10\ \text{mm} \\
\alpha_g &= \text{gage block expansion coefficient} = 11.5 \times 10^{-6}/\text{°C} \\
\alpha_m &= \text{micrometer expansion coefficient} = 5.6 \times 10^{-6}/\text{°C} \\
\Delta T &= \text{ambient temperature} - \text{reference temperature} = 23\ \text{°C} - 20\ \text{°C} = 3\ \text{°C}.
\end{aligned}
$$

Therefore, the change in length is computed as

$$
\begin{aligned}
\Delta x &= 10\ \text{mm} \times (11.5 - 5.6) \times 10^{-6}/\text{°C} \times 3\ \text{°C} \\
&= 1.77 \times 10^{-4}\ \text{mm} \\
&= 0.177\ \mu m.
\end{aligned}
$$

The length measurement can be referenced back to 20 °C by subtracting 0.177 µm from the data sample average. However, we must account for the error in this length correction due to errors in the monitoring temperature and expansion coefficients. The error model is developed as follows:

$$\varepsilon_{\Delta x} = \Delta x - \Delta x_{true} \tag{5-9}$$

where

$$\Delta x_{true} = \left[\left(\alpha_g - \varepsilon_{\alpha_g}\right) - \left(\alpha_m - \varepsilon_{\alpha_m}\right)\right] \times \left[\Delta T - \varepsilon_{\Delta T}\right]. \tag{5-10}$$

Substituting equations (5-8) and (5-10) into equation (5-9) yields

51

$$\varepsilon_{\Delta x} = x_{nom}\Delta T \varepsilon_{\alpha_g} - x_{nom}\Delta T \varepsilon_{\alpha_m} + x_{nom}\alpha_g \varepsilon_{\Delta T} - x_{nom}\alpha_m \varepsilon_{\Delta T}$$
$$- x_{nom}\varepsilon_{\alpha_g}\varepsilon_{\Delta T} + x_{nom}\varepsilon_{\alpha_m}\varepsilon_{\Delta T}.$$

(5-11)

The last two terms in equation (5-11) are referred to as second order terms and are considered to be small compared to the other first order terms. Neglecting second order terms, we can express the length change error equation in a simpler form.

$$\varepsilon_{\Delta x} = x_{nom}\Delta T \varepsilon_{\alpha_g} - x_{nom}\Delta T \varepsilon_{\alpha_m} + x_{nom}\left(\alpha_g - \alpha_m\right)\varepsilon_{\Delta T}$$

(5-12)

The coefficients for ε_{α_g}, ε_{α_m} and $\varepsilon_{\Delta T}$ are actually the partial derivatives of Δx with respect to α_g, α_m and ΔT.

$$\frac{\partial \Delta x}{\partial \alpha_g} = x_{nom}\Delta T = c_g, \quad \frac{\partial \Delta x}{\partial \alpha_m} = -x_{nom}\Delta T = c_m \quad \text{and} \quad \frac{\partial \Delta x}{\partial \Delta T} = x_{nom}\left(\alpha_g - \alpha_m\right) = c_{\Delta T}$$

Therefore, the length change error can be expressed as

$$\varepsilon_{\Delta x} = c_g \varepsilon_{\alpha_g} + c_m \varepsilon_{\alpha_m} + c_{\Delta T}\varepsilon_{\Delta T}$$

(5-13)

where c_g, c_m and $c_{\Delta T}$ are **sensitivity coefficients** that determine the relative contribution of the temperature and expansion coefficient errors to the length change error.

Applying the variance operator to equation (5-13) we have

$$\text{var}\left(\varepsilon_{\Delta x}\right) = \text{var}\left(c_g \varepsilon_{\alpha_g} + c_m \varepsilon_{\alpha_m} + c_{\Delta T}\varepsilon_{\Delta T}\right)$$
$$= c_g^2 \text{var}(\varepsilon_{\alpha_g}) + c_m^2 \text{var}(\varepsilon_{\alpha_m}) + c_{\Delta T}^2 \text{var}(\varepsilon_{\Delta T}) + 2c_g c_m \text{cov}\left(\varepsilon_{\alpha_g},\varepsilon_{\alpha_m}\right)$$
$$+ 2c_g c_{\Delta T} \text{cov}\left(\varepsilon_{\alpha_g},\varepsilon_{\Delta T}\right) + 2c_m c_{\Delta T} \text{cov}\left(\varepsilon_{\alpha_m},\varepsilon_{\Delta T}\right).$$

(5-14)

From Axiom 2, the uncertainty in the length change error can be expressed as

$$u_{\varepsilon_{\Delta x}} = \sqrt{\begin{array}{l} c_g^2 u_{\varepsilon_{\alpha_g}}^2 + c_m^2 u_{\varepsilon_{\alpha_m}}^2 + c_{\Delta T}^2 u_{\varepsilon_{\Delta T}}^2 + 2c_g c_m \rho_{\varepsilon_{\alpha_g},\varepsilon_{\alpha_m}} u_{\varepsilon_{\alpha_g}} u_{\varepsilon_{\alpha_m}} \\ + 2c_g c_{\Delta T}\rho_{\varepsilon_{\alpha_g},\varepsilon_{\Delta T}} u_{\varepsilon_{\alpha_g}} u_{\varepsilon_{\Delta T}} + 2c_m c_{\Delta T}\rho_{\varepsilon_{\alpha_m},\varepsilon_{\Delta T}} u_{\varepsilon_{\alpha_m}} u_{\varepsilon_{\Delta T}} \end{array}}$$

(5-15)

where the last three terms account for any error correlations.

There is no physical reason to believe that a correlation exists between the expansion coefficient errors. Similarly, there shouldn't be any correlation between the temperature error and the expansion coefficient errors. Therefore,

$$\rho_{\varepsilon_{\alpha_g},\varepsilon_{\alpha_m}} = 0, \quad \rho_{\varepsilon_{\alpha_g},\varepsilon_{\Delta T}} = 0, \quad \rho_{\varepsilon_{\alpha_m},\varepsilon_{\Delta T}} = 0$$

and the uncertainty in the length change error can be expressed as

$$u_{\varepsilon_{\Delta x}} = \sqrt{c_g^2 u_{\varepsilon_{\alpha_g}}^2 + c_m^2 u_{\varepsilon_{\alpha_m}}^2 + c_{\Delta T}^2 u_{\varepsilon_{\Delta T}}^2} \ . \tag{5-16}$$

The appropriate probability distribution for the temperature error and expansion coefficient errors is the normal distribution. Therefore, the associated uncertainties can be estimated from the containment limits, containment probability and the inverse normal distribution function. In this analysis, we will assume 95% containment probability for all three error sources.

The uncertainty in the temperature error is estimated from $\pm 2\,°C$ containment limits and a 95% containment probability.

$$u_{\varepsilon_{\Delta T}} = \frac{2}{\Phi^{-1}\left(\dfrac{1+0.95}{2}\right)} = \frac{2}{1.9600} = 1.02\,°C$$

> **Note:** In this example, only the error resulting from the temperature measuring device is considered. However, other error sources resulting from variation in the room temperature and in the gage block and micrometer temperatures during the measurement process may also need to be considered.

The uncertainty in the gage block expansion coefficient is estimated from $\pm 1 \times 10^{-6}/°C$ containment limits and a 95% containment probability.

$$u_{\varepsilon_{\alpha_g}} = \frac{1 \times 10^{-6}\,/°C}{\Phi^{-1}\left(\dfrac{1+0.95}{2}\right)} = \frac{1 \times 10^{-6}\,/°C}{1.9600} = 0.510 \times 10^{-6}\,/°C$$

The uncertainty in the micrometer expansion coefficient is estimated from $\pm 0.5 \times 10^{-6}/°C$ containment limits and a 95% containment probability.

$$u_{\varepsilon_{\alpha_m}} = \frac{0.5 \times 10^{-6}\,/°C}{\Phi^{-1}\left(\dfrac{1+0.95}{2}\right)} = \frac{0.5 \times 10^{-6}\,/°C}{1.9600} = 0.255 \times 10^{-6}\,/°C$$

The corresponding sensitivity coefficients are

$$
\begin{aligned}
c_g &= 10\,\text{mm} \times 3\,°C = 30\,\text{mm-}°C = 3 \times 10^4\ \mu\text{m-}°C \\
c_m &= -10\,\text{mm} \times 3\,°C = -30\,\text{mm-}°C = -3 \times 10^4\ \mu\text{m-}°C \\
c_{\Delta T} &= 10\,\text{mm} \times (11.5 - 5.6) \times 10^{-6}/°C = 5.9 \times 10^{-5}\,\text{mm/}°C = 5.9 \times 10^{-2}\ \mu\text{m/}°C
\end{aligned}
$$

and the uncertainty in the length change error is computed to be

$$u_{\varepsilon_{\Delta x}} = \sqrt{\left(3\times10^4\right)^2 \times \left(0.510\times10^{-6}\right)^2 + \left(-3\times10^4\right)^2 \times \left(0.255\times10^{-6}\right)^2 + \left(5.9\times10^{-2}\right)^2 \times \left(1.020\right)^2}\ \mu m$$

$$= \sqrt{2.34\times10^{-4} + 5.85\times10^{-5} + 3.62\times10^{-3}}\ \mu m$$

$$= \sqrt{3.91\times10^{-3}}\ \mu m = 0.063\ \mu m\,.$$

Thus, the uncertainty due to environmental factors error is

$$u_{env} = u_{\varepsilon_{\Delta x}} = 0.0625\ \mu m\,.$$

5.5   Combine Uncertainties

With the variance addition rule and Axiom 2, we have a method for combining the measurement process uncertainties u_{Gbias}, $u_{\bar{x},ran}$, u_{Mres}, u_{op} and u_{env}. No correlations should exist between measurement process errors, so the uncertainty in the length measurement can be expressed as

$$u_{\varepsilon_{\bar{x}}} = \sqrt{u_{Gbias}^2 + u_{\bar{x},ran}^2 + u_{Mres}^2 + u_{op}^2 + u_{env}^2}\,. \tag{5-17}$$

Therefore, the uncertainty in the average length measurement is computed to be

$$u_{\varepsilon_{\bar{x}}} = \sqrt{(0.042)^2 + (0.463)^2 + (0.289)^2 + (0.304)^2 + (0.063)^2}\ \mu m$$

$$= \sqrt{0.396}\ \mu m = 0.629\ \mu m.$$

The effective degrees of freedom, v_{eff}, for the combined uncertainty can be estimated using the Welch-Satterthwaite formula

$$
v_{eff} = \frac{u_{\varepsilon_{\bar{x}}}^4}{\dfrac{u_{Gbias}^4}{v_{Gbias}} + \dfrac{u_{\bar{x},ran}^4}{v_{\bar{x},ran}} + \dfrac{u_{Mres}^4}{v_{Mres}} + \dfrac{u_{op}^4}{v_{op}} + \dfrac{u_{env}^4}{v_{env}}}
$$

$$
= \frac{u_{\varepsilon_{\bar{x}}}^4}{\dfrac{u_{Gbias}^4}{\infty} + \dfrac{u_{\bar{x},ran}^4}{7} + \dfrac{u_{Mres}^4}{\infty} + \dfrac{u_{op}^4}{\infty} + \dfrac{u_{env}^4}{\infty}} = 7 \times \frac{u_{\varepsilon_{\bar{x}}}^4}{u_{\bar{x},ran}^4}\,. \tag{5-18}
$$

The degrees of freedom for the combined uncertainty are computed to be

$$v_{eff} = 7 \times \frac{\left(0.629\ \mu m\right)^4}{\left(0.463\ \mu m\right)^4} = 7 \times 3.4 = 23.8$$

and are rounded to the nearest whole number, 24.

5.6   Report Analysis Results

All measurement uncertainties relevant to the micrometer calibration process have been taken into account and the analysis results can now be evaluated. In calibration, uncertainty analysis is important for two main reasons. First, to identify excessive uncertainties due to sources of error in our measurement process. Second, to communicate the quantity of interest and its associated uncertainty or to decide whether the quantity is in-tolerance.

5.6.1   Average Measured Value and Combined Uncertainty

As previously stated, the quantity of interest is the average length measurement corrected to 20 °C. In this analysis, the average length measurement at 20 °C is computed to be

$$\bar{x} = 10.003 \text{ mm} - 0.000177 \text{ mm} = 10.0028 \text{ mm}$$

with a combined uncertainty of 0.629 μm with 24 degrees of freedom.

5.6.2   Measurement Process Errors and Uncertainties

The measurement process errors, distributions, uncertainties and degrees of freedom are summarized in Table 5-4. The relative contributions of the measurement process uncertainties to the overall uncertainty in the average length measurement are shown in Figure 5-2. The pareto chart[46] shows that the uncertainties due to repeatability, operator bias, and micrometer resolution are the largest contributors to the 0.629 μm combined uncertainty.

Table 5-4. Measurement Process Uncertainties for Micrometer Calibration

Error Source	Containment Limits	Containment Probability	Error Distribution	Standard Uncertainty	Estimate Type	Deg. of Freedom
Gage Block Bias	+0.18, -0.13	99.00	Lognormal	0.042 μm	B	∞
Repeatability				0.463 μm	A	7
Micrometer Resolution	± 0.5	100.00	Uniform	0.289 μm	B	∞
Operator Bias	± 0.5	90.00	Normal	0.304 μm	B	∞
Environmental Factors	± 0.123	95.00	Normal	0.063 μm	B	∞

---

[46] A Pareto (pronounced puh-RAY-toe) chart is a special type of bar chart where the values plotted are arranged in descending order of importance. The chart is based on the Pareto principle, which states that when several factors affect a situation, a few factors will account for most of the impact.

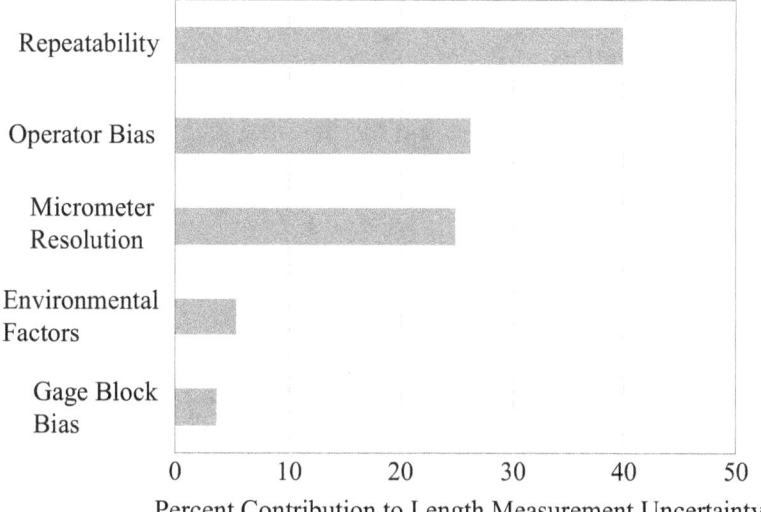

Figure 5-2. Pareto Chart for Micrometer Calibration

5.6.3 Confidence Limits

The combined uncertainty and degrees of freedom can be used to compute confidence limits that are expected to contain the true value, μ, with some specified confidence level or probability, p. The confidence limits are expressed as

$$\bar{x} \pm t_{\alpha/2,\nu} u_{\varepsilon_{\bar{x}}} \tag{5-19}$$

where the multiplier, $t_{\alpha/2\nu}$, is the t-statistic and $\alpha = 1 - p$.

In this analysis, we will use a 95% confidence level (i.e., $p = 0.95$). With a corresponding t-statistic $t_{0.025,24} = 2.0639$, the confidence limits are computed to be

$$10.0028 \text{ mm} \pm 2.0639 \times 0.629 \text{ µm} \quad \text{or} \quad 10.0028 \text{ mm} \pm 1.30 \text{ µm}.$$

5.6.4 In-tolerance Probability

The last step in this analysis example is to determine if the micrometer measurement of the gage block 10 mm nominal length is within the ± 4 µm manufacturer specified tolerance limits. To do this, we must evaluate the micrometer bias, the gage block bias and the uncertainties in these biases.

Recall from equation (5-1), the measured value x is defined by

$$x = x_{true} + \varepsilon_x$$

where

x_{true}	=	true gage block length
ε_x	=	total error in the length measurement.

The nominal gage block length, x_{nom}, is related to the true length by

$$x_{nom} = x_{true} + \varepsilon_{Gbias} \tag{5-20}$$

56

where

$$\varepsilon_{Gbias} = \text{bias in the gage block length.}$$

The difference between the measured value and the nominal gage block length is defined as

$$
\begin{aligned}
\delta &= x - x_{nom} \\
&= \left(x_{true} + \varepsilon_x\right) - \left(x_{true} + \varepsilon_{Gbias}\right) \\
&= \varepsilon_x - \varepsilon_{Gbias}
\end{aligned}
\tag{5-21}
$$

where δ is a measure of the micrometer bias, ε_{Mbias}. Substituting equation (5-3) into equation (5-21), the uncertainty equation for ε_{Mbias} is

$$
\begin{aligned}
u_{\varepsilon_{Mbias}} &= \sqrt{\text{var}\left(\varepsilon_x - \varepsilon_{Gbias}\right)} = \sqrt{\text{var}\left(\varepsilon_{ran} + \varepsilon_{Mres} + \varepsilon_{op} + \varepsilon_{env}\right)} \\
&= \sqrt{u_{ran}^2 + u_{Mres}^2 + u_{op}^2 + u_{env}^2}
\end{aligned}
\tag{5-22}
$$

Replacing u_{ran} in equation (5-22) with $u_{x,ran} = 1.31$, the combined uncertainty is computed to be

$$
\begin{aligned}
u_{\varepsilon_{Mbias}} &= \sqrt{(1.31)^2 + (0.289)^2 + (0.304)^2 + (0.0626)^2} \ \mu\text{m} \\
&= \sqrt{1.896} \ \mu\text{m} = 1.377 \ \mu\text{m}
\end{aligned}
$$

The degrees of freedom for the combined uncertainty is computed to be

$$
\nu_{eff} = 7 \times \frac{u_{\varepsilon_x}^4}{u_{x,ran}^4} = 7 \times \frac{(1.377 \ \mu\text{m})^4}{(1.31 \ \mu\text{m})^4} = 7 \times 1.22 = 8.5
$$

where the value is rounded to the nearest whole number, 9.

The measurement results indicate that the average deviation from the gage block nominal length is $\bar{\delta} = +2.8 \ \mu\text{m}$. The confidence limits for a single value of δ are expressed as

$$
\bar{\delta} \pm t_{\alpha/2,\nu} u_{\varepsilon_{Mbias}} .
\tag{5-23}
$$

For a 95% confidence level, $t_{0.025,9} = 2.2622$ and the confidence limits for δ (e.g., ε_{Mbias}) are computed to be

$$2.8 \ \mu\text{m} \pm 2.2622 \times 1.377 \ \mu\text{m} \ \text{ or } \ 2.8 \ \mu\text{m} \pm 3.12 \ \mu\text{m}$$

Figure 5-3 shows the distribution for ε_{Mbias} relative to the manufacturer specification limits. The shaded area depicts the probability that ε_{Mbias} falls outside of the micrometer specification limits.

There is a much higher probability that the micrometer bias is within the manufacturer specifications than outside them. However, the in-tolerance probability needs to be computed and evaluated to decide whether or not the micrometer's performance is acceptable for its intended application.

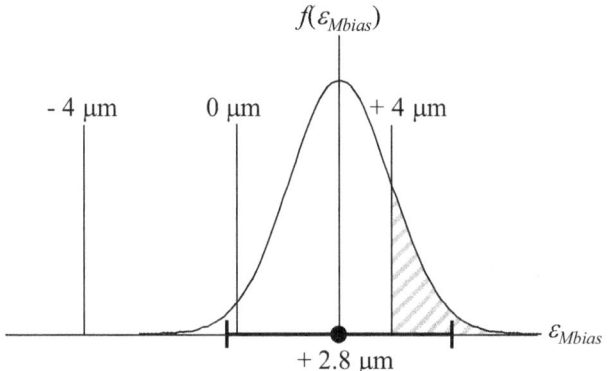

Figure 5-3. Micrometer Bias Distribution

In this decision-making process, it is important to account for the fact that the observed deviation from nominal, δ, is also affected by the bias in the gage block length, ε_{Gbias}. Consequently, the actual micrometer bias may be larger or smaller than δ.

The value of ε_{Gbias} is unknown, but its uncertainty was estimated to be $u_{Gbias} = 0.0423$ μm. This uncertainty is much smaller than the micrometer bias uncertainty, $u_{Mbias} = 1.377$ μm. Therefore, one might deduce that ε_{Gbias} has a minor impact on δ. However, a small value for u_{Gbias} does not preclude a large value for ε_{Gbias}.

To adequately determine micrometer in-tolerance probability, it is also necessary to estimate ε_{Gbias} and the probability that ε_{Gbias} is within its specified tolerance limits. The calculation of biases and in-tolerance probabilities is beyond the scope of this document. Readers are referred to NASA *Measurement Quality Assurance Handbook* Annex 4 – *Estimation and Evaluation of Measurement Decision Risk*.

CHAPTER 6: MULTIVARIATE MEASUREMENTS

This chapter discusses the approach used to estimate the uncertainty of a quantity that is computed from measurements of two or more attributes or parameters. The multivariate uncertainty analysis procedure consists of the following steps:

1. Develop the Parameter Value Equation
2. Develop the Error Model
3. Develop the Uncertainty Model
4. Identify the Measurement Process Errors
5. Estimate Measurement Process Uncertainties
6. Compute Uncertainty Components
7. Account for Cross-Correlations
8. Combine Uncertainty Components
9. Report Analysis Results

The procedure for developing error models and uncertainty models from the parameter value equation is presented. Identifying measurement process errors, estimating their uncertainties and accounting for cross-correlations is also presented. The volume occupied by a cylinder obtained from length and diameter measurements is used to illustrate the concepts and methods of conducting a multivariate uncertainty analysis.

6.1 Develop the Parameter Value Equation

The parameter value equation is a mathematical relationship between the quantity of interest (subject parameter) and the variables or quantities to be measured. The parameter value equation is also referred to as the governing or system equation. For example, consider a case with three measured variables or quantities, x, y, z

$$q = f(x, y, z) \tag{6-1}$$

where

q = subject parameter or quantity of interest
f = mathematical function that relates q to measured quantities x, y, and z.

6.1.1 Cylinder Volume Example

In this analysis example, a steel cylinder artifact with nominal design dimensions of 0.65 cm in length by 1.40 cm in diameter is measured with a micrometer. The objective is to estimate the uncertainty in the cylinder volume measurement.

The parameter value equation for the cylinder volume is given as

$$V = \pi L \left(\frac{D}{2}\right)^2 \tag{6-2}$$

where L and D are the cylinder length and diameter, respectively.

From equation (6-2), we see that, to determine the cylinder volume, we need to measure the length and diameter. The cylinder volume (i.e., parameter value) is then computed based on the values obtained for the length and diameter.

6.2   Develop the Error Model

The error model provides a mathematical relationship between the total error in the quantity of interest to the errors in the measured quantities. The error model is determined from the parameter value equation using a first-order Taylor series approximation.[47]

For example, the error model for ε_q in terms of the error components ε_x, ε_y and ε_y is developed by apply a first-order Taylor Series approximation to equation (6-1).

$$\varepsilon_q = c_x \varepsilon_x + c_y \varepsilon_y + c_z \varepsilon_z \qquad (6\text{-}3)$$

> **Note:** For a multivariate measurement, errors in the measured quantities are called *error components*.

The coefficients, c_x, c_y, and c_z are sensitivity coefficients that determine the relative contribution of the error components to the total error. The sensitivity coefficients are defined as

$$c_x = \frac{\partial f(x,y,z)}{\partial x} = \frac{\partial q}{\partial x} \;,\; c_y = \frac{\partial f(x,y,z)}{\partial y} = \frac{\partial q}{\partial y} \;,\; c_z = \frac{\partial f(x,y,z)}{\partial z} = \frac{\partial q}{\partial z}$$

> **Note:** The sensitivity coefficients are constants computed at a specified set of values for x, y, and z. These may be measured values or other values that are relevant to the measurement process being analyzed.

6.2.1   Cylinder Volume Example

Errors in the length and diameter measurements contribute to the overall error in the estimation of the cylinder volume. In this example, the error model for the cylinder volume equation is developed algebraically to illustrate how the sensitivity coefficients for the length and diameter errors obtained in this manner compare to coefficients obtained using partial derivatives.

By definition,

$$V = V_0 + \varepsilon_V$$
$$D = D_0 + \varepsilon_D$$
$$L = L_0 + \varepsilon_L$$

where

V_0 = nominal or design volume
D_0 = nominal or design diameter
L_0 = nominal or design length

---

[47] Taylor Series, named after English mathematician Brook Taylor, allows the representation of a function as an infinite sum of terms calculated from its derivatives at a specified value. This 1st order approximation is applicable to most measurement scenarios encountered in testing and calibration. However, in the evaluation of stochastic processes, approximations may require the inclusion of 2nd order or higher terms.

$$\varepsilon_V \quad = \quad \text{cylinder volume error}$$
$$\varepsilon_D \quad = \quad \text{diameter measurement error}$$
$$\varepsilon_L \quad = \quad \text{length measurement error}$$

Therefore, equation (6-2) can be expressed as

$$V_0 + \varepsilon_V = \pi \left(\frac{D_0 + \varepsilon_D}{2} \right)^2 (L_0 + \varepsilon_L). \tag{6-4}$$

By rearranging equation (6-4), we obtain an algebraic expression for the cylinder volume error.

$$\begin{aligned}
\varepsilon_V &= \pi \left(\frac{D_0 + \varepsilon_D}{2} \right)^2 (L_0 + \varepsilon_L) - V_0 \\
&= \pi \left(\frac{D_0 + \varepsilon_D}{2} \right)^2 (L_0 + \varepsilon_L) - \pi \left(\frac{D_0}{2} \right)^2 L_0 \\
&= \frac{\pi}{4} \left(D_0^2 + 2 D_0 \varepsilon_D + \varepsilon_D^2 \right) (L_0 + \varepsilon_L) - \frac{\pi}{4} D_0^2 L_0 \\
&= \frac{\pi}{4} \left(D_0^2 L_0 + 2 D_0 L_0 \varepsilon_D + L_0 \varepsilon_D^2 + D_0^2 \varepsilon_L + 2 D_0 \varepsilon_D \varepsilon_L + \varepsilon_D^2 \varepsilon_L \right) - \frac{\pi}{4} D_0^2 L_0
\end{aligned} \tag{6-5}$$

The higher order terms, $L_0 \varepsilon_D^2$, $2 D_0 \varepsilon_D \varepsilon_L$ and $\varepsilon_D^2 \varepsilon_L$, are considered to be small compared to the other first order terms. Neglecting these terms, the cylinder volume error equation can be expressed in a simpler form.

$$\begin{aligned}
\varepsilon_V &= \frac{\pi}{4} \left(D_0^2 L_0 + 2 D_0 L_0 \varepsilon_D + D_0^2 \varepsilon_L \right) - \frac{\pi}{4} D_0^2 L_0 \\
&= \frac{\pi}{4} D_0^2 L_0 - \frac{\pi}{4} D_0^2 L_0 + \frac{\pi}{2} D_0 L_0 \varepsilon_D + \frac{\pi}{4} D_0^2 \varepsilon_L \\
&= \frac{\pi}{2} D_0 L_0 \varepsilon_D + \frac{\pi}{4} D_0^2 \varepsilon_L
\end{aligned} \tag{6-6}$$

Rearranging equation (6-6), yields

$$\varepsilon_V = \pi \left(\frac{D_0}{2} \right)^2 \varepsilon_L + \pi L_0 \frac{D_0}{2} \varepsilon_D. \tag{6-7}$$

The coefficients for ε_L and ε_D in equation (6-7) are actually the partial derivatives of V with respect to L and D.

$$c_L = \frac{\partial V}{\partial L} = \pi \left(\frac{D}{2} \right)^2 \quad \text{and} \quad c_D = \frac{\partial V}{\partial D} = \pi L \left(\frac{D}{2} \right)$$

Therefore, the cylinder volume error model can be expressed as

$$\varepsilon_V = c_L \varepsilon_L + c_D \varepsilon_D \tag{6-8}$$

where the sensitivity coefficients, c_L and c_D, determine the relative contribution of the errors in length and diameter measurements to the total measurement error.

6.3 Develop the Uncertainty Model

As discussed in Chapter 2, the uncertainty in a quantity or variable is the square root of the variable's mean square error or variance. In mathematical terms, this is expressed as

$$u_q = \sqrt{\mathrm{var}(\varepsilon_q)} . \tag{6-9}$$

Applying the variance operator to equation (6-3) gives

$$\begin{aligned} u_q &= \sqrt{\mathrm{var}\left(\varepsilon_q\right)} = \sqrt{\mathrm{var}\left(c_x \varepsilon_x + c_y \varepsilon_y + c_z \varepsilon_z\right)} \\ &= \sqrt{c_x^2 u_x^2 + c_y^2 u_y^2 + c_z^2 u_z^2 + 2 c_x c_y \rho_{xy} u_x u_y + 2 c_x c_z \rho_{xz} u_x u_z + 2 c_y c_z \rho_{yz} u_y u_z} \end{aligned} \tag{6-10}$$

where ρ_{xy}, ρ_{xz} and ρ_{yz} are the correlation coefficients for the errors in x, y and z.

6.3.1 Cylinder Volume Example ....

Applying the variance addition operator to equation (6-8), the uncertainty in the cylinder volume can be expressed as

$$\begin{aligned} u_V &= \sqrt{\mathrm{var}\left(\varepsilon_V\right)} = \sqrt{\mathrm{var}\left(c_L \varepsilon_L + c_D \varepsilon_D\right)} \\ &= \sqrt{c_L^2 u_L^2 + c_D^2 u_D^2 + 2 c_L c_D \rho_{LD} u_L u_D} \end{aligned} \tag{6-11}$$

where ρ_{LD} is the correlation coefficient for the length and diameter errors.

6.4 Identify Measurement Process Errors

As discussed in Chapter 2, measurement process errors are the basic elements of uncertainty analysis. Once these fundamental error sources have been identified, we can begin to develop uncertainty estimates.

6.4.1 Cylinder Volume Example

In this example, the measurement process error sources are:

1. Bias in the micrometer readings (*bias*).
2. Repeatability or random error resulting from different values obtained from measurement to measurement (*ran*).
3. Resolution error due to the finite resolution of the micrometer readings (*res*).
4. Operator bias on the part of the measuring technician (*op*).

5. Error resulting from any thermal or other correction due to a departure from nominal environmental conditions (*env*).

The errors in length and diameter measurements, ε_L and ε_D, can be expressed in terms of their constituent process errors.

$$\varepsilon_L = \varepsilon_{Lbias} + \varepsilon_{Lran} + \varepsilon_{Lres} + \varepsilon_{Lop} + \varepsilon_{Lenv} \qquad (6\text{-}12)$$

and

$$\varepsilon_D = \varepsilon_{Dbias} + \varepsilon_{Dran} + \varepsilon_{Dres} + \varepsilon_{Dop} + \varepsilon_{Denv}. \qquad (6\text{-}13)$$

For this example, the nominal or design specifications for the steel cylinder at 20 °C are

Length (L_0) = 0.65 cm
Diameter (D_0) = 1.40 cm
Volume (V_0) = 1.0 cc

and the measurement process specifications are

Micrometer Bias:	± 0.1mm with 97.5% confidence
Digital Resolution:	0.1 mm
Ambient Temperature:	24 °C ± 2.5 °C with 95% confidence
Thermal Expansion Coefficient for Steel:	5.3×10^{-6} / °C ± 0.5×10^{-6} / °C
Thermal Expansion Coefficient for Micrometer:	1.2×10^{-6} / °C ± 0.2×10^{-6} / °C

Repeat measurements of the cylinder length and diameter, collected in pairs, yielded the data listed in Table 6-1.

Table 6-1. Offset from Nominal Values

Sample Number	Length Offset (mm)	Diameter Offset (mm)
1	0.4	0.2
2	0.3	0.3
3	0.3	0.4
4	0.4	0.5
5	0.5	0.3
6	0.3	0.2
7	0.4	0.4

6.5   Estimate Measurement Process Uncertainties

The specification information and the data in Table 6-1 are used to estimate the process uncertainties for the cylinder length and diameter measurements. The methods of uncertainty estimation are summarized below.

u_{Lbias}, u_{Dbias} - Measurement bias uncertainty is determined heuristically from micrometer tolerance limits and in-tolerance probabilities.

u_{Lran}, u_{Dran} - Repeatability uncertainty is determined statistically from measurement data.

u_{Lres}, u_{Dres} - Resolution uncertainty is determined heuristically from the micrometer resolution specification and containment probability.

u_{Lop}, u_{Dop} - Operator bias uncertainty is determined heuristically based on the micrometer resolution and a containment probability.

u_{Lenv}, u_{Denv} - Environmental factors uncertainty is determined heuristically from tolerances and in-tolerance probabilities for the environment monitoring equipment.

6.5.1 Measurement Bias Uncertainty

Measurement bias can be considered to be a normally distributed error source. Therefore, the uncertainty in the micrometer bias can be expressed in terms of the ± 0.1 mm containment limits, 97.5% containment probability, and the inverse normal distribution function, $\Phi^{-1}(\cdot)$

$$
\begin{aligned}
u_{bias} &= \frac{0.1 \text{ mm}}{\Phi^{-1}\left[(1+0.975)/2\right]} \\
&= \frac{0.1 \text{ mm}}{2.2414} = 0.045 \text{ mm} = 0.0045 \text{ cm}.
\end{aligned}
$$

The micrometer is used to measure cylinder length and diameter, so $u_{Lbias} = u_{Dbias} = 0.0045$ cm.

6.5.2 Repeatability Uncertainty

As discussed in Chapter 3, repeatability uncertainty is equal to the standard deviation of the sample data.

$$
u_{\varepsilon_{x,ran}} = s_x
$$

where

$$
s_x = \sqrt{\frac{1}{n-1}\sum_{i=1}^{n}(x_i - \bar{x})^2}
$$

and x_i is the ith reading and the mean value of the sample is computed from

$$
\bar{x} = \frac{1}{n}\left(x_1 + x_2 + ... + x_n\right).
$$

In this example, the length measurements are recorded in offset units from the nominal length, L_0. The mean of the offset values for the cylinder length is

$$
\begin{aligned}
\bar{L}_{offset} &= \frac{(0.4 + 0.3 + 0.3 + 0.4 + 0.5 + 0.3 + 0.4) \text{ mm}}{7} \\
&= \frac{2.6 \text{ mm}}{7} = 0.37 \text{ mm} = 0.037 \text{ cm}
\end{aligned}
$$

and the differences between the measured offset values and the mean offset value are

$$L_{offset_1} - \bar{L}_{offset} = 0.4 - 0.37 = 0.03 \text{ mm}$$
$$L_{offset_2} - \bar{L}_{offset} = 0.3 - 0.37 = -0.07 \text{ mm}$$
$$L_{offset_3} - \bar{L}_{offset} = 0.3 - 0.37 = -0.07 \text{ mm}$$
$$L_{offset_4} - \bar{L}_{offset} = 0.4 - 0.37 = 0.03 \text{ mm}$$
$$L_{offset_5} - \bar{L}_{offset} = 0.5 - 0.37 = 0.13 \text{ mm}$$
$$L_{offset_6} - \bar{L}_{offset} = 0.3 - 0.37 = -0.07 \text{ mm}$$
$$L_{offset_7} - \bar{L}_{offset} = 0.4 - 0.37 = 0.03 \text{ mm}.$$

The standard deviation is

$$s_{L_{offset}} = \sqrt{\frac{(0.03)^2 + (-0.07)^2 + (-0.07)^2 + (0.03)^2 + (0.13)^2 + (-0.07)^2 + (0.03)^2}{6}} \text{ mm}$$

$$= \sqrt{\frac{0.0343}{6}} \text{ mm} = 0.076 \text{ mm} = 0.0076 \text{ cm}.$$

Thus, the repeatability uncertainty for the cylinder length measurement is

$$u_{Lran} = 0.0076 \text{ cm}.$$

The mean or average cylinder length measurement is

$$\bar{L} = L_0 + \bar{L}_{offset}$$
$$= (0.65 + 0.037) \text{ cm}$$
$$= 0.687 \text{ cm}$$

and the repeatability uncertainty in the mean cylinder length is

$$u_{\bar{L}ran} = \frac{0.0076}{\sqrt{7}} = 0.0029 \text{ cm}.$$

The mean length will be used to compute the cylinder volume, so $u_{\bar{L}ran}$ will be used in the combined uncertainty estimate.

Similarly, the mean of the offset values for the cylinder diameter is

$$\bar{D}_{offset} = \frac{(0.2 + 0.3 + 0.4 + 0.5 + 0.3 + 0.2 + 0.4)}{7} \text{ mm}$$

$$= \frac{2.3}{7} \text{ mm} = 0.33 \text{ mm} = 0.033 \text{ cm}$$

and the differences between the measured offset values and the mean offset value are

$$D_{offset_1} - \bar{D}_{offset} = 0.2 - 0.33 = -0.13 \text{ mm}$$

$$D_{offset_2} - \bar{D}_{offset} = 0.3 - 0.33 = -0.03 \text{ mm}$$

$$D_{offset_3} - \bar{D}_{offset} = 0.4 - 0.33 = 0.07 \text{ mm}$$

$$D_{offset_4} - \bar{D}_{offset} = 0.5 - 0.33 = 0.17 \text{ mm}$$

$$D_{offset_5} - \bar{D}_{offset} = 0.3 - 0.33 = -0.03 \text{ mm}$$

$$D_{offset_6} - \bar{D}_{offset} = 0.2 - 0.33 = -0.13 \text{ mm}$$

$$D_{offset_7} - \bar{D}_{offset} = 0.4 - 0.33 = 0.07 \text{ mm}.$$

The standard deviation is

$$s_{D_{offset}} = \sqrt{\frac{(-0.13)^2 + (-0.03)^2 + (0.07)^2 + (0.17)^2 + (-0.03)^2 + (-0.13)^2 + (0.07)^2}{6}} \text{ mm}$$

$$= \sqrt{\frac{0.0743}{6}} \text{ mm} = 0.11 \text{ mm} = 0.011 \text{ cm}.$$

Thus, repeatability uncertainty for the cylinder diameter measurement is

$$u_{Dran} = 0.011 \text{ cm}.$$

The mean or average cylinder diameter measurement is

$$\bar{D} = D_0 + \bar{D}_{offset}$$

$$= (1.40 + 0.033) \text{ cm}$$

$$= 1.433 \text{ cm}$$

and the repeatability uncertainty in the mean cylinder diameter is

$$u_{\bar{D}ran} = \frac{0.011 \text{ cm}}{\sqrt{7}} = 0.0042 \text{ cm}.$$

The mean diameter will be used to compute the cylinder volume, so $u_{\bar{D}ran}$ will be used in the combined uncertainty estimate.

6.5.3 Resolution Uncertainty

To estimate the resolution uncertainty, we note that the micrometer has a digital readout. Therefore, the resolution error can be assumed to be uniformly distributed with $\pm\,0.05$ mm containment limits and 100% containment probability. Therefore, the resolution uncertainty is computed to be

$$u_{res} = \frac{0.05\,\text{mm}}{\sqrt{3}} = 0.029\ \text{mm} = 0.0029\ \text{cm}$$

Since the micrometer is used to measure cylinder length and diameter,

$$u_{Lres} = u_{Dres} = 0.0029\ \text{cm}.$$

6.5.4 Operator Bias Uncertainty

Operator bias can be considered to be a normally distributed error source. To estimate operator bias uncertainty, we will assume containment limits that are based on roughly half of the resolution error with 90% containment probability. This results in an operator bias uncertainty of

$$u_{op} = \frac{(0.5)(0.01\,\text{cm})}{\Phi^{-1}\big[(1+0.90)/2\big]} = 0.0030\ \text{cm}.$$

The same person measured cylinder length and diameter, so $u_{Lop} = u_{Dop} = 0.0030$ cm.

> **Note:** Containment limits for the operator bias are not necessarily based on resolution error. Any appropriate knowledge about operator bias can be used to develop containment limits and confidence levels.

6.5.5 Environmental Factors Uncertainty

We are interested in determining the uncertainty in the length and diameter measurements resulting from temperature effects. Therefore, we must consider the thermal expansion of the cylinder and the micrometer, as well as the uncertainty in the environmental temperature measurement and the uncertainty in the expansion coefficients.[48]

The effect of temperature deviation from 20 °C on the measured cylinder length is

$$\Delta L = L_0 \times (\alpha_c - \alpha_m) \times \Delta T \tag{6-14}$$

where

α_c = cylinder expansion coefficient = $5.3 \times 10^{-6}/°C$
α_m = micrometer expansion coefficient = $1.2 \times 10^{-6}/°C$
ΔT = ambient temperature – reference temperature = $24\ °C - 20\ °C = 4\ °C$
L_0 = nominal cylinder length = 0.65 cm.

---

[48] This analysis is similar to the environmental factors error model developed in Chapter 5, Section 5.4.5.

Similarly, the effect of temperature deviation from 20 °C on the measured cylinder diameter is

$$\Delta D = D_0 \times (\alpha_c - \alpha_m) \times \Delta T \qquad (6\text{-}15)$$

where D_0 = nominal cylinder diameter = 1.40 cm.

The length change error is expressed as

$$\varepsilon_{\Delta L} = c_{L1}\varepsilon_{\alpha_c} + c_{L2}\varepsilon_{\alpha_m} + c_{L3}\varepsilon_{\Delta T} \qquad (6\text{-}16)$$

where

$$c_{L1} = \frac{\partial \Delta L}{\partial \alpha_c} = L_0 \Delta T, \quad c_{L2} = \frac{\partial \Delta L}{\partial \alpha_m} = -L_0 \Delta T \quad \text{and} \quad c_{L3} = \frac{\partial \Delta L}{\partial \Delta T} = L_0(\alpha_c - \alpha_m).$$

The diameter change error is expressed as

$$\varepsilon_{\Delta D} = c_{D1}\varepsilon_{\alpha_c} + c_{D2}\varepsilon_{\alpha_m} + c_{D3}\varepsilon_{\Delta T} \qquad (6\text{-}17)$$

where

$$c_{D1} = \frac{\partial \Delta D}{\partial \alpha_c} = D_0 \Delta T, \quad c_{D2} = \frac{\partial \Delta D}{\partial \alpha_m} = -D_0 \Delta T \quad \text{and} \quad c_{D3} = \frac{\partial \Delta D}{\partial \Delta T} = D_0(\alpha_c - \alpha_m).$$

Applying the variance operator to equation (6-16) we have

$$
\begin{aligned}
\mathrm{var}\left(\varepsilon_{\Delta L}\right) &= \mathrm{var}\left(c_{L1}\varepsilon_{\alpha_c} + c_{L2}\varepsilon_{\alpha_m} + c_{L3}\varepsilon_{\Delta T}\right) \\
&= c_{L1}^2 \,\mathrm{var}(\varepsilon_{\alpha_c}) + c_{L2}^2 \,\mathrm{var}(\varepsilon_{\alpha_m}) + c_{L3}^2 \,\mathrm{var}(\varepsilon_{\Delta T}) + 2c_{L1}c_{L2}\,\mathrm{cov}\left(\varepsilon_{\alpha_c}, \varepsilon_{\alpha_m}\right) \\
&\quad + 2c_{L1}c_{L3}\,\mathrm{cov}\left(\varepsilon_{\alpha_c}, \varepsilon_{\Delta T}\right) + 2c_{L2}c_{L3}\,\mathrm{cov}\left(\varepsilon_{\alpha_m}, \varepsilon_{\Delta T}\right).
\end{aligned}
\qquad (6\text{-}18)
$$

From Axiom 2, the uncertainty in the length change error can be expressed as

$$u_{\Delta L} = \sqrt{
\begin{aligned}
& c_{L1}^2 u_{\varepsilon_{\alpha_c}}^2 + c_{L2}^2 u_{\varepsilon_{\alpha_m}}^2 + c_{L3}^2 u_{\varepsilon_{\Delta T}}^2 + 2c_{L1}c_{L2}\rho_{\varepsilon_{\alpha_c},\varepsilon_{\alpha_m}} u_{\varepsilon_{\alpha_c}} u_{\varepsilon_{\alpha_m}} \\
& + 2c_{L1}c_{L3}\rho_{\varepsilon_{\alpha_c},\varepsilon_{\Delta T}} u_{\varepsilon_{\alpha_c}} u_{\varepsilon_{\Delta T}} + 2c_{L2}c_{L3}\rho_{\varepsilon_{\alpha_m},\varepsilon_{\Delta T}} u_{\varepsilon_{\alpha_m}} u_{\varepsilon_{\Delta T}}
\end{aligned}
}. \qquad (6\text{-}19)$$

No correlations should exist between the expansion coefficient errors, ε_{α_c} and ε_{α_m}, or between the temperature error, $\varepsilon_{\Delta T}$, and the expansion coefficient errors. Therefore,

$$\rho_{\varepsilon_{\alpha_c},\varepsilon_{\alpha_m}} = 0, \quad \rho_{\varepsilon_{\alpha_c},\varepsilon_{\Delta T}} = 0, \quad \rho_{\varepsilon_{\alpha_m},\varepsilon_{\Delta T}} = 0$$

and the uncertainty in the length change error can be expressed as

$$u_{\Delta L} = \sqrt{c_{L1}^2 u_{\varepsilon_{\alpha_c}}^2 + c_{L2}^2 u_{\varepsilon_{\alpha_m}}^2 + c_{L3}^2 u_{\varepsilon_{\Delta T}}^2} \ . \tag{6-20}$$

Similarly, the uncertainty in the diameter change error can be expressed as

$$u_{\Delta D} = \sqrt{c_{D1}^2 u_{\varepsilon_{\alpha_c}}^2 + c_{D2}^2 u_{\varepsilon_{\alpha_m}}^2 + c_{D3}^2 u_{\varepsilon_{\Delta T}}^2} \ . \tag{6-21}$$

The appropriate probability distribution for the temperature error and expansion coefficient errors is the normal distribution. Therefore, the associated uncertainties can be estimated from the containment limits, containment probability and the inverse normal distribution function. In this analysis, we will use a 95% containment probability for all three error sources.

The uncertainty in the temperature measurement error is expressed in terms of $\pm\,2.5\ °C$ containment limits and 95% containment probability.

$$u_{\varepsilon_{\Delta T}} = \frac{2.5\ °C}{\Phi^{-1}\left[(1+0.95)/2\right]} = \frac{2.5\ °C}{1.9600} = 1.276\ °C$$

Note: In this example, only the error resulting from the temperature measuring device is considered. However, other error sources resulting from variation in the room temperature and in the cylinder and micrometer temperatures during the measurement process may also need to be considered.

The uncertainty in the cylinder expansion coefficient is estimated from $\pm\,0.5 \times 10^{-6}/°C$ containment limits and 95% containment probability.

$$u_{\varepsilon_{\alpha_c}} = \frac{0.5 \times 10^{-6}\ /°C}{\Phi^{-1}\left(\dfrac{1+0.95}{2}\right)} = \frac{0.5 \times 10^{-6}\ /°C}{1.9600} = 0.255 \times 10^{-6}\ /°C$$

The uncertainty in the micrometer expansion coefficient is estimated from $\pm\,0.2 \times 10^{-6}/°C$ containment limits and 95% containment probability.

$$u_{\varepsilon_{\alpha_m}} = \frac{0.2 \times 10^{-6}\ /°C}{\Phi^{-1}\left(\dfrac{1+0.95}{2}\right)} = \frac{0.2 \times 10^{-6}\ /°C}{1.9600} = 0.102 \times 10^{-6}\ /°C$$

The sensitivity coefficients for equation (6-20) are

$$
\begin{aligned}
c_{L1} &= 0.65\ \text{cm} \times 4\ °C = 2.6\ \text{cm-}°C \\
c_{L2} &= -0.65\ \text{cm} \times 4\ °C = -2.6\ \text{cm-}°C \\
c_{L3} &= 0.65\ \text{cm} \times (5.3 - 1.2) \times 10^{-6}/°C = 2.67 \times 10^{-6}\ \text{cm/}°C
\end{aligned}
$$

and the uncertainty in the cylinder length change error is computed to be

$$u_{\Delta L} = \sqrt{(2.6)^2 \times \left(0.255 \times 10^{-6}\right)^2 + (-2.6)^2 \times \left(0.102 \times 10^{-6}\right)^2 + \left(2.67 \times 10^{-6}\right)^2 \times (1.276)^2} \ \text{cm}$$

$$= \sqrt{1.21 \times 10^{-11}} \ \text{cm} = 3.48 \times 10^{-6} \text{cm}.$$

The uncertainty in the cylinder length due to environmental factors error is

$$u_{Lenv} = u_{\Delta L} = 3.48 \times 10^{-6} \ \text{cm}.$$

The sensitivity coefficients for equation (6-21) are

$$
\begin{aligned}
c_{D1} &= 1.40 \ \text{cm} \times 4 \ ^{\circ}\text{C} = 5.6 \ \text{cm-}^{\circ}\text{C} \\
c_{D2} &= -1.40 \ \text{cm} \times 4 \ ^{\circ}\text{C} = -5.6 \ \text{cm-}^{\circ}\text{C} \\
c_{D3} &= 1.40 \ \text{cm} \times (5.3 - 1.2) \times 10^{-6}/^{\circ}\text{C} = 5.74 \times 10^{-6} \ \text{cm/}^{\circ}\text{C}
\end{aligned}
$$

and the uncertainty in the cylinder diameter change error is computed to be

$$u_{\Delta D} = \sqrt{(5.6)^2 \times \left(0.255 \times 10^{-6}\right)^2 + (-5.6)^2 \times \left(0.102 \times 10^{-6}\right)^2 + \left(5.74 \times 10^{-6}\right)^2 \times (1.276)^2} \ \text{cm}$$

$$= \sqrt{5.60 \times 10^{-11}} \ \text{cm}$$

$$= 7.48 \times 10^{-6} \text{cm}.$$

The uncertainty in the cylinder diameter due to environmental factors error is

$$u_{Denv} = u_{\Delta D} = 7.48 \times 10^{-6} \ \text{cm}.$$

6.6 Compute Uncertainty Components

Applying the variance operator to equation (6-12), the uncertainty in the average cylinder length measurement can be expressed as

$$u_{\bar{L}} = \sqrt{u_{Lbias}^2 + u_{Lran}^2 + u_{Lres}^2 + u_{Lop}^2 + u_{Lenv}^2} \ . \tag{6-22}$$

Similarly, applying the variance operator to equation (6-13) gives the following expression for the uncertainty in the average cylinder diameter measurement

$$u_{\bar{D}} = \sqrt{u_{Dbias}^2 + u_{Dran}^2 + u_{Dres}^2 + u_{Dop}^2 + u_{Denv}^2} \ . \tag{6-23}$$

> **Note:** There are no terms correlating process uncertainties within each component expression because the length measurement process errors are independent of one another, as are the diameter measurement process errors.

The uncertainty in the average length measurement is computed to be

$$u_{\bar{L}} = \sqrt{(0.0045\ \text{cm})^2 + (0.0029\ \text{cm})^2 + (0.0029\ \text{cm})^2 + (0.0030\ \text{cm})^2 + (3.48 \times 10^{-6}\ \text{cm})^2}$$

$$= \sqrt{4.61 \times 10^{-5}}\ \text{cm} = 0.0068\ \text{cm}$$

The uncertainty in the average diameter measurement is computed to be

$$u_{\bar{D}} = \sqrt{(0.0045\ \text{cm})^2 + (0.0042\ \text{cm})^2 + (0.0029\ \text{cm})^2 + (0.0030\ \text{cm})^2 + (7.48 \times 10^{-6}\ \text{cm})^2}$$

$$= \sqrt{0.000055\ \text{cm}^2} = 0.0074\ \text{cm}$$

The degrees of freedom for the component uncertainties are computed using the Welch-Satterthwaite formula

$$v_{\bar{L}} = \frac{u_{\bar{L}}^4}{\dfrac{u_{Lbias}^4}{v_{Lbias}} + \dfrac{u_{\bar{L}ran}^4}{v_{\bar{L}ran}} + \dfrac{u_{Lres}^4}{v_{Lres}} + \dfrac{u_{Lop}^4}{v_{Lop}} + \dfrac{u_{Lenv}^4}{v_{Lenv}}} \tag{6-24}$$

and

$$v_{\bar{D}} = \frac{u_{\bar{D}}^4}{\dfrac{u_{Dbias}^4}{v_{Dbias}} + \dfrac{u_{\bar{D}ran}^4}{v_{\bar{D}ran}} + \dfrac{u_{Dres}^4}{v_{Dres}} + \dfrac{u_{Dop}^4}{v_{Dop}} + \dfrac{u_{Denv}^4}{v_{Denv}}}. \tag{6-25}$$

The degrees of freedom for all of the process uncertainties were assumed to be infinite, except for the repeatability uncertainties, $u_{\bar{L}ran}$ and $u_{\bar{D}ran}$, which have degrees of freedom equal to 6 (i.e., sample size minus one). Therefore, the degrees of freedom for the component uncertainties are computed to be

$$v_{\bar{L}} = v_{\bar{L}ran} \times \frac{u_{\bar{L}}^4}{u_{\bar{L}ran}^4} = 6 \times \left(\frac{0.0068\ \text{cm}}{0.0029\ \text{cm}} \right)^4 = 181.4$$

$$v_{\bar{D}} = v_{\bar{D}ran} \times \frac{u_{\bar{D}}^4}{u_{\bar{D}ran}^4} = 6 \times \left(\frac{0.0074\ \text{cm}}{0.0042\ \text{cm}} \right)^4 = 57.8$$

where the degrees of freedom are reported to the nearest whole numbers, $v_{\bar{L}} = 181$ and $v_{\bar{D}} = 58$.

6.7   Account for Cross-Correlations

Before we combine the length and diameter measurement uncertainties, we must consider if there are any cross-correlations between the length and diameter measurement process errors. First, we need to write an equation that expresses the correlation coefficient, ρ_{LD}, for the component errors, ε_L and ε_D, in terms of the cross-correlation coefficients for the process errors

$$\rho_{LD} = \frac{1}{u_L u_D} \sum_{i=1}^{n_i} \sum_{j=1}^{n_j} \rho_{Li,Dj} u_{Li} u_{Dj} \qquad (6\text{-}26)$$

where $\rho_{Li,Dj}$ is the cross-correlation coefficient between the ε_{Li} and ε_{Dj} process errors for the length and diameter components, respectively.

The cross-correlation coefficients can range from minus one to plus one. A positive coefficient applies when the error sources are directly related. A negative coefficient is used when the error sources are inversely related.

Second, let us review what we know about the cylinder measurement process.

1. Both length and diameter are measured using the same device (i.e., a micrometer).

2. All measurements are made by the same person (operator).

3. All measurements were made in the same measuring environment.

Given this knowledge, we can assert that the following process errors are cross-correlated between the length and diameter components:

- Measurement Bias - ε_{Lbias} and ε_{Dbias}
- Operator Bias - ε_{Lop} and ε_{Dop}
- Environmental Factors - ε_{Lenv} and ε_{Denv}

Therefore, equation (6-26) becomes

$$\rho_{LD} = \frac{1}{u_L u_D} \left(\rho_{Lbias,Dbias} u_{Lbias} u_{Dbias} + \rho_{Lop,Dop} u_{Lop} u_{Dop} + \rho_{Lenv,Denv} u_{Lenv} u_{Denv} \right). \qquad (6\text{-}27)$$

6.7.1   Measurement Biases

Since the same device is used to measure the cylinder length and diameter, the micrometer bias for these measurements is the same. In this instance, the cross-correlation coefficient $\rho_{Lbias,Dbias}$ is equal to 1.0.

> **Note:** The micrometer bias may vary slightly over its range. However, in this analysis we assume that this variation is negligible.

6.7.2   Operator Biases

Although the same operator makes both length and diameter measurements, human inconsistency prevents us from assigning a correlation coefficient equal to 1.0. However, we also know that the correlation coefficient should not be equal to zero either. Given that this is all we can say from heuristic considerations, we will set the cross-correlation coefficient between length and diameter operator biases $\rho_{Lop,Dop}$ equal to 0.5.

6.7.3 Environmental Factors Errors

As shown in Section 6.5.5, the length and diameter change errors, $\varepsilon_{\Delta L}$ and $\varepsilon_{\Delta D}$, are functions of the expansion coefficient and temperature change errors. Consequently, an increase or decrease in $\varepsilon_{\Delta L}$ will result in a proportionate increase or decrease in $\varepsilon_{\Delta D}$. Therefore, the cross-correlation coefficient $\rho_{Lenv,Denv}$, is equal to 1.0.

The correlation coefficient ρ_{LD} can now be expressed as

$$\rho_{LD} = \frac{1}{u_L u_D}\left(u_{Lbias}u_{Dbias} + 0.5u_{Lop}u_{Dop} + u_{Lenv}u_{Denv}\right). \tag{6-28}$$

6.8 Combine Uncertainty Components

The equation for the cylinder volume uncertainty is obtained by substituting equation (6-28) into equation (6-11)

$$u_{\bar{V}} = \sqrt{c_{\bar{L}}^2 u_{\bar{L}}^2 + c_{\bar{D}}^2 u_{\bar{D}}^2 + 2c_{\bar{L}}c_{\bar{D}}\left(u_{Lbias}u_{Dbias} + 0.5u_{Lop}u_{Dop} + u_{Lenv}u_{Denv}\right)} \tag{6-29}$$

where the sensitivity coefficients are

$$c_{\bar{L}} = \pi\left(\frac{\bar{D}}{2}\right)^2 = 3.14159\left(\frac{1.433 \text{ cm}}{2}\right)^2 = 1.613 \text{ cm}^2$$

and

$$c_{\bar{D}} = \pi\bar{L}\left(\frac{\bar{D}}{2}\right) = 3.14159 \times 0.687 \text{ cm} \times \left(\frac{1.433 \text{ cm}}{2}\right) = 1.547 \text{ cm}^2.$$

The cylinder volume uncertainty is computed to be

$$u_{\bar{V}} = \sqrt{\begin{array}{l}\left(1.613 \text{ cm}^2\right)^2\left(0.0068 \text{ cm}\right)^2 + \left(1.547 \text{ cm}^2\right)^2\left(0.0074 \text{ cm}\right)^2 \\ +2\left(1.613 \text{ cm}^2\right)\left(1.547 \text{ cm}^2\right)\left[\left(0.0045 \text{ cm}\right)^2 + 0.5\left(0.003 \text{ cm}\right)^2 + \left(3.48 \text{ cm}\right)\left(7.48 \text{ cm}\right) \times 10^{-12}\right]\end{array}}$$

$$= \sqrt{\left(1.20 \text{ cm}^6 + 1.31 \text{ cm}^6 + 1.24 \text{ cm}^6\right) \times 10^{-4}}$$

$$= \sqrt{3.75 \times 10^{-2}} \text{ cm}^3 = 0.0194 \text{ cm}^3.$$

The degrees of freedom for the cylinder volume uncertainty are estimated using the Welch-Satterthwaite formula

$$v = \frac{u_{\bar{V}*}^4}{\dfrac{c_{\bar{L}}^4 u_{\bar{L}}^4}{v_{\bar{L}}} + \dfrac{c_{\bar{D}}^4 u_{\bar{D}}^4}{v_{\bar{D}}}} \tag{6-30}$$

where $u_{\bar{V}*}$ is the total uncertainty computed without cross-correlations between the uncertainty components $u_{\bar{L}}$ and $u_{\bar{D}}$.[49]

$$u_{\bar{V}*} = \sqrt{c_{\bar{L}}^2 u_{\bar{L}}^2 + c_{\bar{D}}^2 u_{\bar{D}}^2} = \sqrt{\left(1.613 \text{ cm}^2 \times 0.0068 \text{ cm}\right)^2 + \left(1.547 \text{ cm}^2 \times 0.0074 \text{ cm}\right)^2}$$

$$= \sqrt{0.00012 \text{ cm}^6 + 0.00013 \text{ cm}^6} = \sqrt{0.00025 \text{ cm}^6} = 0.0158 \text{ cm}^3.$$

The degrees of freedom for the cylinder volume uncertainty are computed to be

$$v_{\bar{V}} = \frac{\left(0.0158 \text{ cm}^3\right)^4}{\dfrac{\left(1.613 \text{ cm}^2 \times 0.0068 \text{ cm}\right)^4}{181} + \dfrac{\left(1.547 \text{ cm}^2 \times 0.0074 \text{ cm}\right)^4}{58}} = 165.7$$

and are reported as the nearest whole number, $v_{\bar{V}} = 166$.

6.9   Report Analysis Results

We have accounted for all uncertainties considered to be relevant to the cylinder volume measurement process and can now evaluate the results of our analysis. In this case, we are interested in the uncertainty in the cylinder volume computed from the average length and diameter measurements corrected to 20 °C.

6.9.1   Cylinder Volume and Combined Uncertainty

The cylinder volume is computed using the average cylinder length and diameter corrected to 20 °C. The average cylinder length and diameter at 24 °C were computed to be 0.687 cm and 1.433 cm, respectively. Equations (6-14) and (6-15) can be used to estimate the effect of temperature deviation from 20 °C on the measured cylinder length and diameter.

$$\begin{aligned}\Delta L &= 0.65 \text{ cm} \times (5.3 - 1.2) \, 10^{-6}/°C \times 4 \, °C \\ &= 1.07 \times 10^{-5} \text{ cm}\end{aligned}$$

$$\begin{aligned}\Delta D &= 1.40 \text{ cm} \times (5.3 - 1.2) \, 10^{-6}/°C \times 4 \, °C \\ &= 2.30 \times 10^{-5} \text{ cm}\end{aligned}$$

Both the length and diameter expansion are considered to be insignificant for this analysis. Therefore, the cylinder volume can be computed using the uncorrected average length and diameter.

$$\bar{V} = \pi\bar{L}\left(\frac{\bar{D}}{2}\right)^2 \tag{6-31}$$

where $\bar{L}$ = 0.687 cm and $\bar{D}$ = 1.433 cm. The cylinder volume is computed to be

---

[49] While the Welch-Satterthwaite formula is applicable for statistically independent, normally distributed error sources it can usually be thought of as a fair approximation in cases where error sources are not statistically independent.

$$\bar{V} = 3.14159 \times 0.687 \times (1.433/2)^2 = 1.108 \text{ cm}^3$$

with an uncertainty of $u_{\bar{V}} = 0.019$ cm^3 and 166 degrees of freedom.

6.9.2 Measurement Process Errors and Uncertainties

The measurement process errors, corresponding distributions, uncertainties and degrees of freedom are summarized in Table 6-2.

Table 6-2. Measurement Process Uncertainties for Cylinder Volume Measurement

Error Source	Error Limits (cm)	Error Containment Probability	Error Distribution	Estimated Standard Uncertainty (cm)	Estimate Type	Deg. of Freed.	Sensitivity Coeff. (cm^2)	Component Uncertainty (cm^3)
ε_{Lbias}	± 0.01	97.5%	Normal	0.0045	B	∞	1.613	0.0073
ε_{Dbias}	± 0.01	97.5%	Normal	0.0045	B	∞	1.547	0.0070
$\varepsilon_{\overline{Lran}}$				0.0029	A	6	1.613	0.0047
$\varepsilon_{\overline{Dran}}$				0.0042	A	6	1.547	0.0065
ε_{Lres}	± 0.01	100%	Uniform	0.0029	B	∞	1.613	0.0047
ε_{Dres}	± 0.01	100%	Uniform	0.0029	B	∞	1.547	0.0045
ε_{Lop}	± 0.01	90%	Normal	0.0030	B	∞	1.613	0.0048
ε_{Dop}	± 0.01	90%	Normal	0.0030	B	∞	1.547	0.0046
ε_{Lenv}			Normal	3.48×10^{-6}	B	∞	1.613	5.61×10^{-6}
ε_{Denv}			Normal	7.48×10^{-6}	B	∞	1.547	1.16×10^{-5}

The component uncertainty is the product of the standard uncertainty and the sensitivity coefficient. The relative contributions of the component uncertainties to the overall cylinder volume uncertainty are shown in Figure 6-1. Recall from equation (6-29), the uncertainty in the cylinder volume accounts for cross-correlations between ε_{Lbias} and ε_{Dbias}, ε_{Lop} and ε_{Dop}, and ε_{Lenv} and ε_{Denv}. Consequently, measurement bias uncertainty (i.e., micrometer bias uncertainty) for length and diameter are the largest contributors to the uncertainty in cylinder volume, followed by operator bias and diameter repeatability.

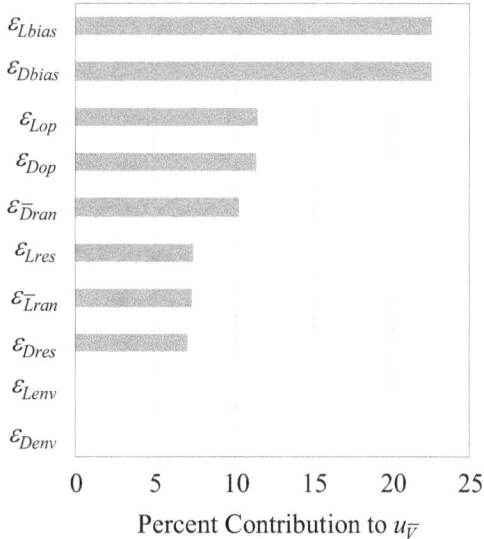

Figure 6-1. Pareto Chart for Cylinder Volume Measurement

6.9.3 Confidence Limits

The combined uncertainty and degrees of freedom can be used to compute confidence limits that are expected to contain the true cylinder volume with some specified confidence level or probability, p. The confidence limits are expressed as

$$\bar{V} \pm t_{\alpha/2,\nu} u_{\bar{V}} \tag{6-32}$$

where the multiplier, $t_{\alpha/2\nu}$, is the t-statistic and $\alpha = 1 - p$.

For this analysis, let us assume that we want 99% confidence limits (i.e., $p = 0.99$). The corresponding t-statistic is $t_{0.005,166} \cong 2.6$ and the confidence limits are computed to be

$$1.108 \text{ cm}^3 \pm 2.6 \times 0.019 \text{ cm}^3 \text{ or } 1.108 \text{ cm}^3 \pm 0.049 \text{ cm}^3.$$

6.9.3.1 Single Cylinder Volume Measurement

To compute the confidence limits for the cylinder volume determined from a single pair of length and diameter measurements, $u_{\bar{L},ran}$ and $u_{\bar{D},ran}$ must be replaced with $u_{L,ran}$ and $u_{D,ran}$ in equations (6-22) and (6-23), respectively.

The uncertainty components, u_L and u_D are then computed to be

$$u_L = \sqrt{\left(0.0045 \text{ cm}\right)^2 + \left(0.0076 \text{ cm}\right)^2 + \left(0.0029 \text{ cm}\right)^2 + \left(0.0030 \text{ cm}\right)^2 + \left(3.48 \text{ cm} \times 10^{-6}\right)^2}$$

$$= \sqrt{9.54 \times 10^{-5} \text{ cm}^2} = 0.0098 \text{ cm}$$

$$u_D = \sqrt{\left(0.0045 \text{ cm}\right)^2 + \left(0.011 \text{ cm}\right)^2 + \left(0.0029 \text{ cm}\right)^2 + \left(0.0030 \text{ cm}\right)^2 + \left(7.48 \text{ cm} \times 10^{-6}\right)^2}$$

$$= \sqrt{1.59 \times 10^{-4} \text{ cm}^2} = 0.0126 \text{ cm}$$

The associated degrees of freedom for these uncertainty components are similarly computed by substituting $u_{\bar{L},ran}$ and $u_{\bar{D},ran}$ with $u_{L,ran}$ and $u_{D,ran}$, respectively.

$$V_L = V_{Lran} \times \frac{u_L^4}{u_{Lran}^4} = 6 \times \left(\frac{0.0098 \text{ cm}}{0.0076 \text{ cm}}\right)^4 = 16.6$$

$$V_D = V_{Dran} \times \frac{u_D^4}{u_{Dran}^4} = 6 \times \left(\frac{0.0126 \text{ cm}}{0.0110 \text{ cm}}\right)^4 = 10.3$$

The degrees of are reported to the nearest whole numbers, $v_L = 17$ and $v_D = 10$.

The cylinder volume uncertainty is then computed by substituting u_L and u_D for $u_{\bar{L}}$ and $u_{\bar{D}}$ in equation (6-29).

$$u_V = \sqrt{\begin{array}{l}\left(1.613 \text{ cm}^2\right)^2 \left(0.0098 \text{ cm}\right)^2 + \left(1.547 \text{ cm}^2\right)^2 \left(0.0126 \text{ cm}\right)^2 \\ +2\left(1.613 \text{ cm}^2\right)\left(1.547 \text{ cm}^2\right)\left[\left(0.0045 \text{ cm}\right)^2 + 0.5\left(0.003 \text{ cm}\right)^2 + \left(3.48 \text{ cm}\right)\left(7.48 \text{ cm}\right) \times 10^{-12}\right]\end{array}}$$

$$= \sqrt{\left(2.48 \text{ cm}^6 + 3.80 \text{ cm}^6 + 1.24 \text{ cm}^6\right) \times 10^{-4}} = \sqrt{7.52} \times 10^{-2} \text{ cm}^3 = 0.027 \text{ cm}^3.$$

The corresponding degrees of freedom are computed using the Welch Satterthwaite formula

$$v = \frac{u_{V*}^4}{\dfrac{c_L^4 u_L^4}{v_L} + \dfrac{c_D^4 u_D^4}{v_D}}$$

where u_{V*} is the cylinder volume uncertainty computed without cross-correlations.

$$u_{V*} = \sqrt{c_L^2 u_L^2 + c_D^2 u_D^2}$$

$$= \sqrt{\left(1.613 \text{ cm}^2 \times 0.0098 \text{ cm}\right)^2 + \left(1.547 \text{ cm}^2 \times 0.0126 \text{ cm}\right)^2}$$

$$= 0.025 \text{ cm}^3.$$

The degrees of freedom for the cylinder volume uncertainty are computed to be

$$v_V = \frac{\left(0.025\ \text{cm}^3\right)^4}{\dfrac{\left(1.613\ \text{cm}^2 \times 0.0098\ \text{cm}\right)^4}{17} + \dfrac{\left(1.547\ \text{cm}^2 \times 0.0126\ \text{cm}\right)^4}{10}} = 21.8\ .$$

and are reported as the nearest whole number, $v_V = 22$.

The confidence limits, relative to a single cylinder volume measurement are

$$\bar{V} \pm t_{\alpha/2, v} u_V\ . \tag{6-33}$$

For a 99% confidence level, $t_{0.005,22} \cong 2.82$ and the confidence limits are computed to be

$$1.108\ \text{cm}^3 \pm 2.82 \times 0.027\ \text{cm}^3\ \text{ or }\ 1.108\ \text{cm}^3 \pm 0.076\ \text{cm}^3\ .$$

CHAPTER 7: MEASUREMENT SYSTEMS

7.0   General

This chapter discusses the approach used to estimate the uncertainty of a quantity (or subject parameter) that is measured with a system comprised of component modules arranged in series. The analysis process traces system uncertainty module by module from system input to system output.

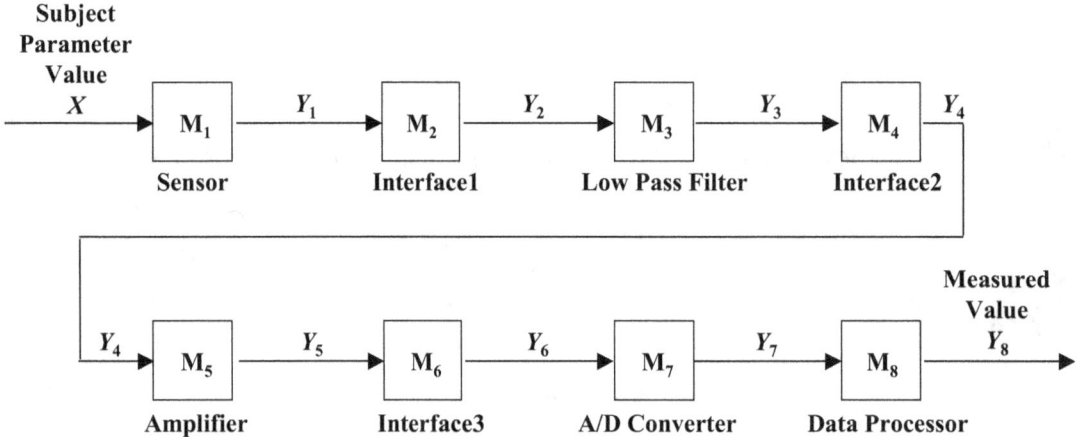

Figure 7-1. Block Diagram for Example System

System uncertainty analysis follows a structured procedure. This is necessary because the output from any given module of a system may comprise the input to another module or modules. Since each module's output carries with it an element of uncertainty, this means that the same uncertainty may be present at the input of some other module.

7.1   System Analysis Procedure

In analyzing linear measurement systems, we develop output equations for each module. From these equations, we identify sources of error for each module. We then estimate the uncertainty in each error source and compute the combined uncertainty in the output of each module. In doing this, we make certain that *the uncertainty in the output of each module is included in the input to the succeeding module in the system.*

In this respect, the system analysis results are computed somewhat differently than those previously discussed for direct measurements and multivariate measurements. The general system analysis procedure consists of the following steps:

1. Develop the System Model
2. Define the System Input
3. Define the System Modules
4. Identify Module Error Sources
5. Develop Module Error Models
6. Develop Module Uncertainty Models
7. Estimate Module Uncertainties

8. Compute System Output Uncertainty
9. Report Analysis Results

The processes for developing a system model and the corresponding module output equations are presented. Processes for identifying measurement process errors, estimating their uncertainties and accounting for correlations are presented using a load cell measurement system for illustration.

7.2   Develop the System Model

The first step in the system analysis procedure is to develop a model that describes the modules involved in processing the measurement of interest (i.e., subject parameter). The model should include a diagram depicting the modules of the system and their inputs and outputs and identify the hardware and software used.

The system diagram can be a useful guide for developing the equations that describe the module outputs in terms of inputs and identify the parameters that characterize these processes. It may also be beneficial to develop a functional model that relates component errors to the overall system output error.

7.2.1   Load Cell Measurement System

In this example, a load cell is calibrated using a weight standard, as illustrated in Figure 7-2. The calibration weight is extended from the load cell via a monofilament line. The DC voltage output from the amplifier module is measured with a digital multimeter (DMM). Three repeat measurements of DC voltage are obtained by adding and removing the calibration weight.

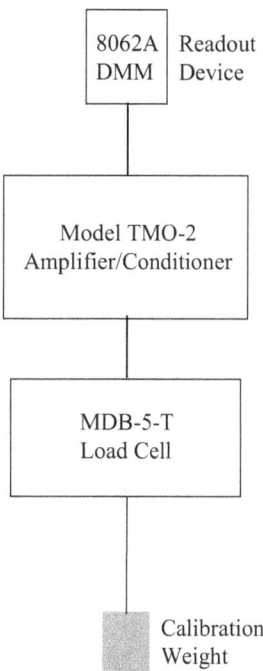

Figure 7-2. Load Cell Calibration Setup

The purpose of this analysis is to estimate and report the total uncertainty in the average DC voltage obtained via the load cell calibration process. For the load cell system analysis, we need to define the mathematical relationship between the quantity being investigated and its

80

component variables. In this case, measurement is made through a linear sequence of stages as shown in Figure 7-3.

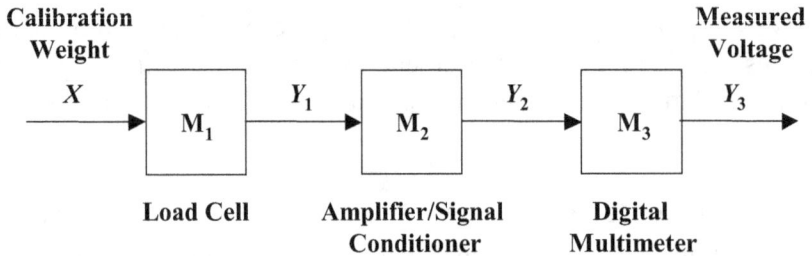

Figure 7-3. Block Diagram of Load Cell Measurement System

The output, *Y*, from any given module of the system may comprise the input of another module or modules. Since each module's output carries with it an element of uncertainty, then this uncertainty may be present at the input of a subsequent module.

7.3   Define the System Input

The second step in the system analysis procedure is to define the quantity or parameter value that is sought through measurement.[50] The nominal (or expected) input value, measurement area and units are specified during this step.

7.3.1   Load Cell Measurement System

As previously indicated, a weight standard is used to calibrate the load cell measurement system. The nominal value of the calibration weight is stated to be 3 lb$_f$. In this case, the nominal value for the system input is 3, the input measurement area is force and the units are lb$_f$.

7.4   Define the System Modules

Once a sufficiently detailed block diagram has been established, the equations that relate the inputs and outputs for each module can be developed. The basic approach is to clearly describe the physical processes that transform the system input along its path from module to module.

7.4.1   Load Cell Module (M₁)

The first module in the load cell measurement system consists of an MDB-5-T load cell manufactured by Transducer Techniques, Inc. This load cell is a passive sensor that requires an external voltage source and has a rated output of 2 mV/V nominal for loads up to 5 lb$_f$. Therefore, the nominal sensitivity of the load cell is 0.4 mV/V/lb$_f$.

The basic transfer function for the load cell module is given in equation (7-1).

$$LC_{Out} = W \times S \times V_{ex} \tag{7-1}$$

where

$$
\begin{aligned}
LC_{Out} &= \text{Load cell output, mV} \\
W &= \text{Applied load or weight, lb}_f \\
S &= \text{Load cell sensitivity, mV/V/lb}_f \\
V_{ex} &= \text{Excitation voltage, V}
\end{aligned}
$$

---

[50] i.e., the input stimulus to the measurement system.

7.4.2 Amplifier Module (M₂)

The second system module is a TMO-2 Amplifier, manufactured by Transducer Techniques Inc. This module amplifies the mV output from the load cell module to V. The nominal amplifier gain is the ratio of the maximum amplifier output to the maximum load cell output. The basic transfer function for this module is given in equation (7-2).

$$Amp_{Out} = LC_{Out} \times G \tag{7-2}$$

where

$$Amp_{Out} = \text{Amplifier Output, V}$$
$$G = \text{Amplifier Gain, V/mV}$$

7.4.3 Digital Multimeter Module (M₃)

An 8602A digital multimeter, manufactured by Fluke, converts the analog output signal from the amplifier module to a digital signal and displays it on a readout device. The basic transfer function for this module is expressed in equation (7-3).

$$DMM_{Out} = Amp_{Out} \tag{7-3}$$

where

$$DMM_{Out} = \text{Digital multimeter output, V}$$

7.5 Identify Module Error Sources

The next analysis step is to evaluate module functions or parameters to identify errors that may contribute to the total module output error.

In the analysis of the load cell measurement system, error in the mass of the calibration weight, errors intrinsic to the measurement equipment used, and other process errors are considered. A list of applicable error sources is given below.

- Bias in the value of the calibration weight
- Errors associated with the MDB-5-T Load Cell
- Errors associated with the Model TMO-2 Amplifier
- Errors associated with the 8062A Digital Multimeter
- Error associated with the repeat measurements taken

7.5.1 Load Cell Module (M₁)

For this module, the following error sources must be considered:

- Bias in the value of the calibration weight
- Excitation voltage error
- Load cell error

Manufacturer's published specifications for the load cell[51] are listed in Table 7-1. The following sources of load cell error will be included:

---

[51] Specifications obtained from www.ttloadcells.com/mdb-load-cell.cfm

- Nonlinearity
- Hysteresis
- Noise
- Zero balance
- Temperature effect on output
- Temperature effect on zero

Table 7-1. MDB-5-T Load Cell Specifications

Specification	Value	Units
Maximum Applied Load	5	lb_f
Rated Output (R.O.)	2	mV/V
Nonlinearity	0.05% of R.O.	mV/V
Hysteresis	0.05% of R.O.	mV/V
Noise (Nonrepeatability)	0.05% of R.O.	mV/V
Zero Balance	1.0% of R.O.	mV/V
Compensated Temp. Range	60 to 160	°F
Temperature Effect on Output	0.005% of Load/°F	lb_f/°F
Temperature Effect on Zero	0.005% of R.O./°F	mV/V/°F
Recommended Excitation Voltage	10	VDC

When developing an equation for the load cell module, the impact of the error sources on the output must be considered. Each of the error sources listed above are discussed briefly to determine how they should be accounted for in the load cell output equation.

7.5.1.1   Calibration Weight

The 3 lb_f calibration weight has specified error limits of ± 0.003 lb_f. In this analysis, these limits are interpreted to represent 99 % confidence limits. The associated error distribution is characterized by the normal distribution.

7.5.1.2   Excitation Voltage

Since the MDB-5-T load cell is a passive sensor, it requires an external power supply. The TMO-2 Amplifier provides a regulated 8 VDC excitation power supply with ± 0.25 V error limits. The excitation voltage error limits are interpreted to be 95% confidence limits for a normally distributed error.

7.5.1.3   Nonlinearity.

Nonlinearity is a measure of the deviation of the actual input-to-output performance of the device from an ideal linear relationship. Nonlinearity error is fixed at any given input, but varies with magnitude and sign over a range of inputs. Therefore, it is considered to be a random error that is normally distributed. The manufacturer specification limits of ± 0.05% of the rated output are interpreted to be a 95% confidence limits.

7.5.1.4   Hysteresis

Hysteresis indicates that the output of the device is dependent upon the direction and magnitude by which the input is changed. At any input value, hysteresis can be expressed as the difference between the ascending and descending outputs. Hysteresis error is fixed at any given input, but varies with magnitude and sign over a range of inputs. Therefore, it is considered to be a random error that is normally distributed. The manufacturer specification limits of ± 0.05% of the rated output are interpreted to be a 95% confidence limits.

7.5.1.5   Noise

Nonrepeatability or random error intrinsic to the device, which causes the output to vary from observation to observation for a constant input is usually specified as noise. This error source varies with magnitude and sign over a range of inputs and is normally distributed. The manufacturer specification limits of ± 0.05% of the rated output are interpreted to be 95% confidence limits.

7.5.1.6   Zero Balance

Zero balance refers to the zero offset that occurs if the device exhibits a non-zero output for a zero input. Although zero offset error can be reduced by adjustment, there is no way to completely eliminate it because we do not know the true value of the offset. The manufacturer specification limits of ± 1% of the rated output are interpreted to be 95% confidence limits for a normally distributed error.

7.5.1.7   Temperature Effects

The load cell is part of a tension testing machine, which heats up during use. The load cell temperature is monitored and recorded during the testing process and observed to increase from 75 °F to 85 °F. The load cell is subjected to the same temperature change during calibration.

Temperature can affect both the offset and sensitivity of the load cell. To establish these effects, the device is typically tested at several temperatures within its operating range and the effects on zero and sensitivity or output are observed.

Although the load cell is used within its compensated temperature range, the manufacturer acknowledges that some compensation error exists, hence the stated specifications for Temperature Effect on Output and Temperature Effect on Zero.

The temperature effect on output of 0.005% load/°F specified by the manufacturer is equivalent to 0.00015 lb_f/°F for an applied load of 3 lb_f. The temperature effect on zero and the temperature effect on output specifications are interpreted to be a 95% confidence limits for normally distributed errors.

A 10 °F temperature change is used in this analysis to account for temperature compensation error. The temperature measurement error limits are ± 2 °F with an associated 99% confidence level. The temperature error is assumed to be normally distributed.

7.5.2   Amplifier Module (M$_2$)

For this module, the following error sources must be considered:

- Load cell output error
- Amplifier error

The manufacturer's published specifications for the amplifier[52] are listed in Table 7-2. For a recommended applied excitation voltage of 10 VDC, the MDB-5-T load cell has a maximum rated output of 20 mV. Therefore, the TMO-2 amplifier has a nominal gain of 10V/20 mV or 0.5 V/mV.

Table 7-2. TMO-2 Amplifier Specifications

Specification	Value	Units
Maximum Output Voltage	10	V
Gain (nominal)	0.5	V/mV
Gain Accuracy	0.05% of Full Scale	mV
Gain Stability	0.01%	mV
Nonlinearity	0.01%	mV
Noise and Ripple	< 3	mV
Balance Stability	0.2%	mV
Temperature Coefficient	0.02% of F.S./°C	mV/°C

Given the above specifications, the following sources of amplifier error are applicable to this analysis:

- Gain accuracy
- Gain stability (or Instability)
- Nonlinearity
- Noise
- Balance stability
- Temperature coefficient

7.5.2.1   Gain Accuracy

Gain is the ratio of the amplifier output signal voltage to the input signal voltage. In this case, the TMO-2 amplifier has a nominal gain of 10V/20 mV or 0.5 V/mV. The manufacturer specified accuracy limits of $\pm$ 0.05% of full scale are interpreted to be 95% confidence limits for a normally distributed error.

7.5.2.2   Gain Stability

If the amplifier voltage gain is represented by G_V, its input resistance by R and its feedback resistance by R_f, then oscillations are possible when

$$\frac{RG_V}{R + R_f} = \pi \,.$$

These oscillations appear as an instability in the amplifier gain. The manufacturer specification of 0.01% is interpreted to be $\pm$ 0.01% of full scale. These limits are assumed to represent 95% confidence limits for a normally distributed error.

---

[52] Specifications obtained from www.ttloadcells.com/TMO-2.cfm

7.5.2.3 Nonlinearity

As with the load cell module, actual amplifier response may depart from the ideal or assumed output versus input curve. Nonlinearity errors are point-by-point differences in actual versus expected response over the range of input signal levels. The manufacturer specification of 0.01% is interpreted to be $\pm$ 0.01% of full scale and representative of 95% confidence limits for a normally distributed errors.

7.5.2.4 Noise

Noise generated within the amplifier that enters the signal path causes errors in the amplifier output. Since noise is directly related to gain, manufacturers usually specify noise error in absolute units of Volts RMS or Volts peak-to-peak. The manufacturer specification of 3 mV peak-to-peak is estimated to be $\pm$ 1.5 mV limits that are equivalent to 99% confidence limits for a normally distributed error.

7.5.2.5 Balance Stability

Balance stability, or instability, refers to a non-zero amplifier output exhibited for a zero input. Although balance instability can be reduced by adjustment, there is no way to completely eliminate it because we do not know the true value of the zero offset. The manufacturer specification of $\pm$ 0.2% is interpreted to be $\pm$ 0.2% of full scale. These limits are also interpreted to be 95% confidence limits for a normally distributed error.

7.5.2.6 Temperature Coefficient

Both the balance (or zero) and gain are affected by temperature. Manufacturers generally state this as a temperature coefficient (or Tempco) in terms of percent change or full scale per degree. The manufacturer specification limits of $\pm$ 0.02% of full scale/°C are interpreted to be 95% confidence limits for a normally distributed error.

To quantify the effect of temperature, however, we must establish the expected temperature change and use this with the temperature coefficient to compute expected variations. As with the load cell module, the impact of temperature correction error is estimated using a temperature range of 5.6 °C (10 °F) with measurement error limits of $\pm$ 1.1 °C with an associated confidence level of 99% for a normally distributed error.

7.5.3 Digital Multimeter Module (M₃)

Manufacturer's published specifications for the DC voltage function of the digital multimeter[53] are listed in Table 7-3. In this module, key error sources include:

- Amplifier output error
- DC voltmeter accuracy
- DC voltmeter digital resolution
- Repeat measurements error

---

[53] Specifications from 8062A Instruction Manual downloaded from www.fluke.com

Table 7-3. 8062A DC Voltage Specifications

Specification	Value	Units
200 mV Range Resolution	0.01	mV
200 mV Range Accuracy	0.05% of Reading + 2 digits	mV
2 V Range Resolution	0.1	mV
2 V Range Accuracy	0.05% of Reading + 2 digits	mV
20 V Range Resolution	1	mV
20 V Range Accuracy	0.07% of Reading + 2 digits	mV

7.5.3.1   DC Voltage Accuracy.

The overall accuracy of the DC Voltage reading for a 20 V range is specified as ± (0.07% of reading + 2 digits). These specification limits are interpreted to be 95% confidence limits for a normally distributed error.

7.5.3.2   Digital Resolution.

The digital resolution for the 20 V DC range is specified as 1 mV. Since this is a digital display, the resolution error is uniformly distributed. Therefore, the resolution error limits ± 0.5 mV are interpreted to be the minimum 100% containment or bounding limits.

7.5.3.3   Repeatability.

Random error resulting from repeat measurements can result from various physical phenomena such as temperature variation or the act of removing and re-suspending the calibration weight multiple times. Repeatability uncertainty will be estimated using the data listed in Table 7-4.

Table 7-4. DC Voltage Readings

Repeat Measurement	Measured DC Voltage (V)	Offset from Nominal DC Voltage (V)
1	4.856	0.056
2	4.861	0.061
3	4.860	0.060

7.6   Develop Module Error Models

The next analysis step is to develop an error model for each module. In most instances, the module output is a function of several variables. Therefore, the error model must be developed using a multivariate analysis approach.

As discussed in Chapter 6, the error model for a multivariate parameter $q = f(x,y,z)$ is expressed as

$$\varepsilon_q = c_x \varepsilon_x + c_y \varepsilon_y + c_z \varepsilon_z ,$$

where c_x, c_y, and c_z are sensitivity coefficients that determine the relative contribution of the errors in x, y and z to the total error in q. The sensitivity coefficients are defined as

$$c_x = \left(\frac{\partial q}{\partial x}\right), \quad c_y = \left(\frac{\partial q}{\partial y}\right), \quad c_z = \left(\frac{\partial q}{\partial z}\right).$$

For the load cell measurement system, equations (7-1) through (7-3) provide the basis for the development of the module error models.

7.6.1 Load Cell Module (M_1)

The load cell output equation (7-1) must be modified before the associated error model can be developed. It is a good practice to first assign names to the relevant module error sources and other parameters. The load cell error source and parameter names, descriptions, nominal values, error limits and confidence levels are listed in Table 7-5.

Table 7-5. Parameters used in Modified Load Cell Module Equation

Parameter Name	Description	Nominal or Mean Value	Error Limits	Percent Confid.
W_C	Calibration Weight or Load	3 lb$_f$	± 0.003 lb$_f$	99
S	Load Cell Sensitivity	0.4 mV/V/lb$_f$		
NL	Nonlinearity	0 mV/V	± 0.001 mV/V	95
Hys	Hysteresis	0 mV/V	± 0.001 mV/V	95
NS	Nonrepeatability	0 mV/V	± 0.001 mV/V	95
ZO	Zero Balance	0 mV/V	± 0.02 mV/V	95
$TR_{\circ F}$	Temperature Range	10 °F	± 2.0 °F	99
TE_{Out}	Temperature Effect on Output	0 lb$_f$/°F	± 1.5 e-4 lb$_f$/°F	95
TE_{Zero}	Temperature Effect on Zero	0 mV/V /°F	± 0.0001 mV/V /°F	95
V_{ex}	Applied Excitation Voltage	8 V	± 0.25 V	95

Next, given what is known about the load cell error sources listed in Table 7-5, they must be appropriately incorporated into equation (7-1). The modified module output equation is given in equation (7-4).

$$LC_{out} = [(W_C + TE_{out} \times TR_{\circ F}) \times S + NL + Hys + NS + ZO + TE_{Zero} \times TR_{\circ F}] \times V_{ex} \qquad (7\text{-}4)$$

From equation (7-4), the error model for the load cell module is given in equation (7-5).

$$
\begin{aligned}
\varepsilon_{LC_{Out}} = {} & c_{W_C}\varepsilon_{W_C} + c_S\varepsilon_S + c_{NL}\varepsilon_{NL} + c_{Hys}\varepsilon_{Hys} + c_{NS}\varepsilon_{NS} + c_{ZO}\varepsilon_{ZO} \\
& + c_{TE_{Out}}\varepsilon_{TE_{Out}} + c_{TE_{Zero}}\varepsilon_{TE_{Zero}} + c_{TR_{\circ F}}\varepsilon_{TR_{\circ F}} + c_{V_{ex}}\varepsilon_{V_{ex}}
\end{aligned}
\qquad (7\text{-}5)
$$

The partial derivative equations used to compute the sensitivity coefficients are listed below.

$$c_{W_C} = \frac{\partial LC_{Out}}{\partial W_C} = S \times V_{ex} \qquad\qquad c_S = \frac{\partial LC_{Out}}{\partial S} = \left(W_C + TE_{out} \times TR_{\circ F}\right) \times V_{ex}$$

$$c_{NL} = \frac{\partial LC_{Out}}{\partial NL} = V_{ex} \qquad\qquad c_{Hys} = \frac{\partial LC_{Out}}{\partial Hys} = V_{ex}$$

$$c_{NS} = \frac{\partial LC_{Out}}{\partial NS} = V_{ex} \qquad\qquad c_{ZO} = \frac{\partial LC_{Out}}{\partial ZO} = V_{ex}$$

$$c_{TE_{Out}} = \frac{\partial LC_{Out}}{\partial TE_{Out}} = TR_{\circ F} \times S \times V_{ex} \qquad c_{TE_{Zero}} = \frac{\partial LC_{Out}}{\partial TE_{Zero}} = TR_{\circ F} \times V_{ex}$$

$$c_{TR_{\circ F}} = \frac{\partial LC_{Out}}{\partial TR_{\circ F}} = (TE_{out} \times S + TE_{zero}) \times V_{ex}$$

$$c_{V_{ex}} = \frac{\partial LC_{Out}}{\partial V_{ex}} = (W_C + TE_{Out} \times TR_{\circ F}) \times S + NL + Hys + NS + ZO + TE_{Zero} \times TR_{\circ F}$$

7.6.2   Amplifier Module (M₂)

The amplifier output equation (7-2) must be modified before the associated error model can be developed. The amplifier error source and parameter names, descriptions, nominal values, error limits and confidence levels are listed in Table 7-6.

Table 7-6. Parameters used in Modified Amplifier Module Equation

Parameter Name	Description	Nominal or Mean Value	Error Limits	Percent Confidence
LC_{Out}	Amplifier Input			
G	Gain	0.5 V/mV		
G_{Acc}	Gain Accuracy	0 V	± 5 mV	95
G_S	Gain Stability	0 V	± 1 mV	95
G_{NL}	Nonlinearity	0 V	± 1 mV	95
G_{NS}	Noise	0 V	± 1.5 mV	99
B_{St}	Balance Stability	0 V	± 20 mV	95
TC	Temperature Coefficient	0 V/°C	± 2 mV/°C	95
$TR_{\circ C}$	Temperature Range	5.6 °C	± 1.1 °C	99

Given what is known about the amplifier error sources listed in Table 7-6, they must be adequately incorporated into the amplifier module output equation (7-2). The modified module output equation is given in equation (7-6).

$$Amp_{Out} = LC_{Out} \times G + G_{Acc} + G_S + G_{NL} + G_{NS} + B_{St} + TC \times TR_{\circ C} \qquad (7\text{-}6)$$

From equation (7-6), the error model for the amplifier module is given in equation (7-7).

$$\begin{aligned}\varepsilon_{Amp_{Out}} = {}& c_{LC_{Out}}\varepsilon_{LC_{Out}} + c_G\varepsilon_G + c_{G_{Acc}}\varepsilon_{G_{Acc}} + c_{G_S}\varepsilon_{G_S} + c_{G_{NL}}\varepsilon_{G_{NL}} \\ & + c_{G_{NS}}\varepsilon_{G_{NS}} + c_{B_{St}}\varepsilon_{B_{St}} + c_{TC}\varepsilon_{TC} + c_{TR_{\circ C}}\varepsilon_{TR_{\circ C}}\end{aligned} \qquad (7\text{-}7)$$

The partial derivative equations used to compute the sensitivity coefficients are listed below.

$$c_{LC_{Out}} = \frac{\partial Amp_{Out}}{\partial LC_{Out}} = G \qquad c_G = \frac{\partial Amp_{Out}}{\partial G} = LC_{Out} \qquad c_{G_{Acc}} = \frac{\partial Amp_{Out}}{\partial G_{Acc}} = 1$$

$$c_{G_S} = \frac{\partial Amp_{Out}}{\partial G_S} = 1 \qquad c_{G_{NL}} = \frac{\partial Amp_{Out}}{\partial G_{NL}} = 1 \qquad c_{G_{NS}} = \frac{\partial Amp_{Out}}{\partial G_{NS}} = 1$$

$$c_{B_{St}} = \frac{\partial Amp_{Out}}{\partial B_{St}} = 1 \qquad c_{TC} = \frac{\partial Amp_{Out}}{\partial TC} = TR_{\circ C} \qquad c_{TR_{\circ C}} = \frac{\partial Amp_{Out}}{\partial TR_{\circ C}} = TC$$

7.6.3 Digital Multimeter Module (M$_3$)

The digital multimeter output equation must also be modified before the associated error model can be developed. The modified multimeter output equation given in equation (7-8) accounts for the relevant module parameters and error limits listed in Table 7-7. The repeatability parameter, V_{ran}, is estimated from the three repeat voltages listed in Table 7-4.

$$DMM_{Out} = Amp_{Out} + DMM_{Acc} + DMM_{res} + V_{ran} \qquad (7\text{-}8)$$

Table 7-7. Parameters used in Modified Multimeter Module Equation

Parameter Name	Description	Nominal or Mean Value	Error Limits	Percent Confidence
Amp_{Out}	DMM Input	4.80 V		
DMM_{Acc}	DC Voltmeter Accuracy	0 V	$\pm$ (0.07% Read + 2 mV)	95
DMM_{res}	DC Voltmeter Digital Resolution	0 V	$\pm$ 0.5 mV	100

The corresponding error model for the multimeter module is given in equation (7-9).

$$\varepsilon_{DMM_{Out}} = c_{Amp_{Out}}\varepsilon_{Amp_{Out}} + c_{DMM_{Acc}}\varepsilon_{DMM_{Acc}} + c_{DMM_{res}}\varepsilon_{DMM_{res}} + c_{V_{ran}}\varepsilon_{V_{ran}} \qquad (7\text{-}9)$$

The partial derivative equations used to compute the sensitivity coefficients are listed below.

$$c_{Amp_{Out}} = \frac{\partial DMM_{Out}}{\partial Amp_{Out}} = 1 \qquad c_{DMM_{Acc}} = \frac{\partial DMM_{Out}}{\partial DMM_{Acc}} = 1$$

$$c_{DMM_{res}} = \frac{\partial DMM_{Out}}{\partial DMM_{res}} = 1 \qquad c_{V_{ran}} = \frac{\partial DMM_{Out}}{\partial V_{ran}} = 1$$

7.7 Develop Module Uncertainty Models

The next step in the system analysis procedure is to develop an uncertainty model for each system module, accounting for correlations between error sources.

As discussed in Chapter 6, the uncertainty in a multivariate parameter q can be determined by applying the variance addition operator

$$u_q = \sqrt{\mathrm{var}\left(c_x\varepsilon_x + c_y\varepsilon_y + c_z\varepsilon_z\right)}$$
$$= \sqrt{c_x^2 u_x^2 + c_y^2 u_y^2 + c_z^2 u_z^2 + 2c_x c_y \rho_{xy} u_x u_y + 2c_x c_z \rho_{xz} u_x u_z + 2c_y c_z \rho_{yz} u_y u_z}$$

where ρ_{xy}, ρ_{xz} and ρ_{yz} are the correlation coefficients for the errors in x, y and z.

7.7.1   Load Cell Module (M$_1$)

The uncertainty model for the load cell module output can be determined by applying the variance operator to equation (7-5).

$$
\begin{aligned}
u_{LC_{Out}} &= \sqrt{\operatorname{var}\left(\varepsilon_{LC_{Out}}\right)} \\
&= \sqrt{\operatorname{var}\left(\begin{array}{l} c_{W_C}\varepsilon_{W_C} + c_S\varepsilon_S + c_{NL}\varepsilon_{NL} + c_{Hys}\varepsilon_{Hys} + c_{NS}\varepsilon_{NS} + c_{ZO}\varepsilon_{ZO} \\ + c_{TE_{Out}}\varepsilon_{TE_{Out}} + c_{TE_{Zero}}\varepsilon_{TE_{Zero}} + c_{TR_{\circ F}}\varepsilon_{TR_{\circ F}} + c_{V_{ex}}\varepsilon_{V_{ex}} \end{array}\right)}
\end{aligned}
\tag{7-10}
$$

There are no correlations between error sources for the load cell module. Therefore, the uncertainty in the load cell output can be expressed as

$$
u_{LC_{Out}} = \sqrt{\begin{array}{l} c_{W_C}^2 u_{W_C}^2 + c_S^2 u_S^2 + c_{NL}^2 u_{NL}^2 + c_{Hys}^2 u_{Hys}^2 + c_{NS}^2 u_{NS}^2 + c_{ZO}^2 u_{ZO}^2 \\ + c_{TE_{Out}}^2 u_{TE_{Out}}^2 + c_{TE_{Zero}}^2 u_{TE_{Zero}}^2 + c_{TR_{\circ F}}^2 u_{TR_{\circ F}}^2 + c_{V_{ex}}^2 u_{V_{ex}}^2 \end{array}}
\tag{7-11}
$$

7.7.2   Amplifier Module (M$_2$)

The uncertainty model for the amplifier module output is developed by applying the variance operator to the corresponding error model given in equation (7-7).

$$
\begin{aligned}
u_{Amp_{Out}} &= \sqrt{\operatorname{var}\left(\varepsilon_{Amp_{Out}}\right)} \\
&= \sqrt{\operatorname{var}\left(\begin{array}{l} c_{LC_{Out}}\varepsilon_{LC_{Out}} + c_G\varepsilon_G + c_{G_{Acc}}\varepsilon_{G_{Acc}} + c_{G_S}\varepsilon_{G_S} + c_{G_{NL}}\varepsilon_{G_{NL}} \\ + c_{G_{NS}}\varepsilon_{G_{NS}} + c_{B_{St}}\varepsilon_{B_{St}} + c_{TC}\varepsilon_{TC} + c_{TR_{\circ C}}\varepsilon_{TR_{\circ C}} \end{array}\right)}
\end{aligned}
\tag{7-12}
$$

There are no correlations between error sources. Therefore, the uncertainty model for the amplifier module output can be expressed as

$$
u_{Amp_{Out}} = \sqrt{\begin{array}{l} c_{LC_{Out}}^2 u_{LC_{Out}}^2 + c_G^2 u_G^2 + c_{G_{Acc}}^2 u_{G_{Acc}}^2 + c_{G_S}^2 u_{G_S}^2 + c_{G_{NL}}^2 u_{G_{NL}}^2 \\ + c_{G_{NS}}^2 u_{G_{NS}}^2 + c_{B_{St}}^2 u_{B_{St}}^2 + c_{TC}^2 u_{TC}^2 + c_{TR_{\circ C}}^2 u_{TR_{\circ C}}^2 \end{array}}
\tag{7-13}
$$

7.7.3   Digital Multimeter Module (M$_3$)

The uncertainty model for the multimeter module output is developed by applying the variance operator to the corresponding error model given in equation (7-9).

$$
u_{DMM_{Out}} = \sqrt{\operatorname{var}\left(\varepsilon_{DMM_{Out}}\right)} = \sqrt{\operatorname{var}\left(\begin{array}{l} c_{Amp_{Out}}\varepsilon_{Amp_{Out}} + c_{DMM_{Acc}}\varepsilon_{DMM_{Acc}} \\ + c_{DMM_{res}}\varepsilon_{DMM_{res}} + c_{V_{ran}}\varepsilon_{V_{ran}} \end{array}\right)}
\tag{7-14}
$$

There are no correlations between error sources and the correlation coefficients all have values of unity. Therefore, the uncertainty model for the multimeter module output can be expressed as

$$u_{DMM_{Out}} = \sqrt{u^2_{Amp_{Out}} + u^2_{DMM_{Acc}} + u^2_{DMM_{res}} + u^2_{V_{ran}}}$$

(7-15)

7.8 Estimate Module Uncertainties

The next step in the system analysis procedure is to estimate uncertainties in module parameters and to use these estimates to compute the combined uncertainty and associated degrees of freedom for each module output.

7.8.1 Load Cell Module (M₁)

The load cell output uncertainty is computed from the uncertainty estimates and sensitivity coefficients for each module parameter.

As discussed in section 7.5.1, all of the error sources identified for the load cell module are assumed to follow a normal distribution. Therefore, the corresponding uncertainties can be estimated from the error limits, $\pm L$, confidence level, p, and the inverse normal distribution function, $\Phi^{-1}(\cdot)$, as discussed in Chapter 3.

$$u = \frac{L}{\Phi^{-1}\left(\dfrac{1+p}{2}\right)}$$

For example, the bias uncertainty of the calibration weight is estimated to be

$$u_{WC} = \frac{0.003 \text{ lb}_f}{\Phi^{-1}\left(\dfrac{1+0.99}{2}\right)} = \frac{0.003 \text{ lb}_f}{2.5758} = 0.0012 \text{ lb}_f.$$

Similarly, the uncertainty due to the excitation voltage error is estimated to be

$$u_{V_{ex}} = \frac{0.25 \text{ V}}{\Phi^{-1}\left(\dfrac{1+0.95}{2}\right)} = \frac{0.25 \text{ V}}{1.9600} = 0.1276 \text{ V}.$$

The sensitivity coefficients are computed using the parameter nominal or mean values.

$$c_{WC} = S \times V_{ex} \qquad\qquad c_S = \left(W_C + TE_{out} \times TR\text{\textdegree}F\right) \times V_{ex}$$
$$= 0.4 \text{ mV/V/lb}_f \times 8 \text{ V} \qquad = \left(3 \text{ lb}_f + 0 \text{ lb}_f /\text{\textdegree}F \times 10 \text{ \textdegree}F\right) \times 8 \text{ V}$$
$$= 3.2 \text{ mV/lb}_f \qquad\qquad = 3 \text{ lb}_f \times 8 \text{ V} = 24 \text{ lb}_f \bullet \text{V}$$

$$c_{NL} = V_{ex} = 8 \text{ V} \qquad c_{Hys} = V_{ex} = 8 \text{ V} \qquad c_{NS} = V_{ex} = 8 \text{ V} \qquad c_{ZO} = V_{ex} = 8 \text{ V}$$

92

$$c_{TR\circ_F} = (TE_{out} \times S + TE_{zero}) \times V_{ex} \qquad c_{TE_{Out}} = TR\circ_F \times S \times V_{ex}$$
$$= (0 \times 0.4 \text{ mV/V/lb}_f + 0) \times 8 \text{ V} \qquad = 10\,°\text{F} \times 0.4 \text{ mV/V/lb}_f \times 8 \text{ V}$$
$$= 0 \qquad\qquad = 32\,°\text{F} \times \text{mV/lb}_f$$

$$c_{TE_{Zero}} = TR\circ_F \times V_{ex}$$
$$= 10\,°\text{F} \times 8 \text{ V}$$
$$= 80\,°\text{F} \times \text{V}$$

$$c_{V_{ex}} = (W_C + TE_{Out} \times TR\circ_F) \times S + NL + Hys + NS + ZO + TE_{Zero} \times TR\circ_F$$
$$= (3 \text{ lb}_f + 0 \text{ lb}_f/°\text{F} \times 10\,°\text{F}) \times 0.4 \text{ mV/V/lb}_f + 0 \text{ mV/V} + 0 \text{ mV/V} + 0 \text{ mV/V}$$
$$+ 0 \text{ mV/V} + 0 \text{ mV/V/°F} \times 10\,°\text{F}$$
$$= 3 \text{ lb}_f \times 0.4 \text{ mV/V/lb}_f = 1.2 \text{ mV/V}$$

The estimated uncertainties and sensitivity coefficients for each parameter are listed in Table 7-8.

Table 7-8. Estimated Uncertainties for Load Cell Module Parameters

Param. Name	Nominal or Mean Value	± Error Limits	Percent Conf.	Standard Uncertainty	Sensitivity Coefficient	Component Uncertainty
W_C	3 lb$_f$	± 0.003 lb$_f$	99	0.0012 lb$_f$	3.2 mV/lb$_f$	0.0037 mV
S	0.4 mV/V/lb$_f$				24 lb$_f$×V	
NL	0 mV/V	± 0.001 mV/V	95	0.0005 mV/V	8 V	0.0041 mV
Hys	0 mV/V	± 0.001 mV/V	95	0.0005 mV/V	8 V	0.0041 mV
NS	0 mV/V	± 0.001 mV/V	95	0.0005 mV/V	8 V	0.0041 mV
ZO	0 mV/V	± 0.02 mV/V	95	0.0102 mV/V	8 V	0.0816 mV
$TR\circ_F$	10 °F	± 2.0 °F	99	0.7764 °F	0	
TE_{Out}	0 lb$_f$/°F	± 1.5 × 10^{-4} lb$_f$/°F	95	0.0001 lb/°F	32 °F×mV/lb$_f$	0.0024 mV
TE_{Zero}	0 mV/°F	± 0.0001 mV/V/°F	95	0.00005 mV/V/°F	80 °F×V	0.0041 mV
V_{ex}	8 V	± 0.25 V	95	0.1276 V	1.2 mV/V	0.1531 mV

The component uncertainties listed in Table 7-8 are the products of the standard uncertainty and sensitivity coefficient for each parameter. From equation (7-1), the nominal load cell output is computed to be

$$LC_{Out} = W \times S \times V_{ex} = 3 \text{ lb}_f \times 0.4 \text{ mV/V/lb}_f \times 8 \text{ V} = 9.60 \text{ mV}.$$

The load cell output uncertainty is computed by taking the root sum square of the component uncertainties.

$$u_{LC_{Out}} = \sqrt{\begin{array}{l}(0.0037 \text{ mV})^2 + (0.0041 \text{ mV})^2 + (0.0041 \text{ mV})^2 + (0.0041 \text{ mV})^2 \\ + (0.0816 \text{ mV})^2 + (0.0024 \text{ mV})^2 + (0.0041 \text{ mV})^2 + (0.1531 \text{ mV})^2\end{array}}$$
$$= \sqrt{0.0302 \text{ mV}^2} = 0.174 \text{ mV}$$

The Welch-Satterthwaite formula given in equation (7-16) is used to compute the degrees of freedom for the load cell output uncertainty.

$$v_{LC_{Out}} = \cfrac{u_{LC_{Out}}^4}{\left[\begin{array}{c} \cfrac{c_{W_C}^4 u_{W_C}^4}{v_{W_C}} + \cfrac{c_{NL}^4 u_{NL}^4}{v_{NL}} + \cfrac{c_{Hys}^4 u_{Hys}^4}{v_{Hys}} + \cfrac{c_{NS}^4 u_{NS}^4}{v_{NS}} + \cfrac{c_{ZO}^4 u_{ZO}^4}{v_{ZO}} + \cfrac{c_{TR\circ_F}^4 u_{TR\circ_F}^4}{v_{TR\circ_F}} \\[2ex] + \cfrac{c_{TE_{Out}}^4 u_{TE_{Out}}^4}{v_{TE_{Out}}} + \cfrac{c_{TE_{Zero}}^4 u_{TE_{Zero}}^4}{v_{TE_{Zero}}} + \cfrac{c_{V_{ex}}^4 u_{V_{ex}}^4}{v_{V_{ex}}} \end{array}\right]} \qquad (7\text{-}16)$$

The degrees of freedom for all of the error source uncertainties are assumed infinite. Therefore, the degrees of freedom for the load cell output uncertainty are also infinite.

7.8.2   Amplifier Module (M₂)

The amplifier output uncertainty is computed from the uncertainty estimates and sensitivity coefficients for each module parameter.

As discussed in section 7.5.2, all of the error sources identified for the amplifier module are assumed to follow a normal distribution. Therefore, the corresponding uncertainties can be estimated from the error limits, confidence level, and the inverse normal distribution function.

For example, the uncertainty due to the gain accuracy is estimated to be

$$u_{G_{Acc}} = \frac{5\,\text{mV}}{\Phi^{-1}\left(\dfrac{1+0.95}{2}\right)} = \frac{5\,\text{mV}}{1.9600} = 2.551\,\text{mV}.$$

The sensitivity coefficients are computed using the parameter nominal or mean values.

$$c_{LC_{Out}} = G = 0.5\,\text{V/mV} \qquad c_G = LC_{Out} = 9.6\,\text{mV} \qquad c_{G_{Acc}} = 1$$

$$c_{G_S} = 1 \qquad c_{G_{NL}} = 1 \qquad c_{G_{NS}} = 1$$

$$c_{B_{St}} = 1 \qquad c_{TC} = TR\circ_C = 5.6\,°C \qquad c_{TR\circ_C} = TC = 0$$

The estimated uncertainties and sensitivity coefficients for each parameter are listed in Table 7-9.

Table 7-9. Estimated Uncertainties for Amplifier Module Parameters

Param. Name	Nominal or Mean Value	± Error Limits	Percent Confid.	Standard Uncertainty	Sensitivity Coefficient	Component Uncertainty
LC_{Out}	9.6 mV			0.1740 mV	0.5 V/mV	0.0869 V
G	0.5 V/mV				9.6 mV	
G_{Acc}	0 V	± 5 mV	95	2.551 mV	1	0.0026 V
G_S	0 V	± 1 mV	95	0.510 mV	1	0.0005 V

94

G_{NL}	0 V	± 1 mV	95	0.510 mV	1	0.0005 V
G_{NS}	0 V	± 1.5 mV	99	0.583 mV	1	0.0006 V
B_{St}	0 V	± 20 mV	95	10.204 mV	1	0.0102 V
TC	0 V	± 2 mV/°C	95	1.020 mV/°C	5.6 °C	0.0057 V
$TR_{°C}$	5.6 °C	± 1.1 °C	99	0.427°C	0	0 V

From equation (7-2), the nominal amplifier output is computed to be

$$Amp_{Out} = LC_{Out} \times G = 9.60 \text{ mV} \times 0.5 \text{ V/mV} = 4.80 \text{ V}.$$

The amplifier output uncertainty is computed by taking the root sum square of the component uncertainties.

$$u_{Amp_{Out}} = \sqrt{\begin{array}{l}(0.0869 \text{ V})^2 + (0.0026 \text{ V})^2 + (0.0005 \text{ V})^2 + (0.0005 \text{ V})^2 \\ + (0.0006 \text{ V})^2 + (0.0102 \text{ V})^2 + (0.0057 \text{ V})^2\end{array}}$$

$$= \sqrt{0.0077 \text{ V}^2} = 0.0877 \text{ V}.$$

The degrees of freedom for the amplifier output uncertainty are computed using the Welch-Satterthwaite formula, as shown in equation (7-17).

$$v_{Amp_{Out}} = \frac{u_{Amp_{Out}}^4}{\left[\begin{array}{l}\dfrac{c_{LC_{Out}}^4 u_{LC_{Out}}^4}{v_{LC_{Out}}} + \dfrac{c_{G_{Acc}}^4 u_{G_{Acc}}^4}{v_{G_{Acc}}} + \dfrac{c_{GS}^4 u_{GS}^4}{v_{GS}} + \dfrac{c_{GNL}^4 u_{GNL}^4}{v_{GNL}} + \dfrac{c_{GNS}^4 u_{GNS}^4}{v_{GNS}} \\ + \dfrac{c_{BST}^4 u_{BST}^4}{v_{BST}} + \dfrac{c_{TC}^4 u_{TC}^4}{v_{TC}} + \dfrac{c_{TR_{°C}}^4 u_{TR_{°C}}^4}{v_{TR_{°C}}}\end{array}\right]} \tag{7-17}$$

The degrees of freedom for all of the error source uncertainties are assumed infinite. Therefore, the degrees of freedom for the amplifier output uncertainty are also infinite.

7.8.3   Multimeter Module (M₃)

The multimeter output uncertainty is computed from the uncertainty estimates and sensitivity coefficients for each module parameter.

As discussed in section 7.5.3, the DMM accuracy error follows a normal distribution. Therefore, the uncertainty due to the digital multimeter accuracy is estimated to be

$$u_{DMM_{Acc}} = \frac{\left(4.8 \text{ V} \times \dfrac{0.07}{100} + 2 \text{ mV} \times \dfrac{1 \text{ V}}{1000 \text{ mV}}\right)}{\Phi^{-1}\left(\dfrac{1+0.95}{2}\right)} = \left(\frac{0.0034 \text{ V} + 0.002 \text{ V}}{1.9600}\right) = \frac{0.0054 \text{ V}}{1.9600} = 0.0027 \text{ V}.$$

The DMM resolution error follows a uniform distribution, so the digital multimeter resolution uncertainty is estimated to be

$$u_{DMM_{res}} = \frac{0.5 \text{ mV}}{\sqrt{3}} = \frac{0.5 \text{ mV}}{1.732} = 0.3 \text{ mV} = 0.0003 \text{ V}.$$

The repeatability uncertainty is the standard deviation of the repeat measurements listed in Table 7-4. The mean voltage offset is

$$\overline{V}_{offset} = \frac{0.056 + 0.061 + 0.060}{3} \text{ V} = \frac{0.177}{3} \text{ V} = 0.059 \text{ V}$$

The differences between the individual voltage offsets and the mean value are

$$V_{1offset} - \overline{V}_{offset} = 0.056 \text{ V} - 0.059 \text{V} = -0.003 \text{V}$$
$$V_{2offset} - \overline{V}_{offset} = 0.061 \text{ V} - 0.059 \text{V} = 0.002 \text{V}$$
$$V_{3offset} - \overline{V}_{offset} = 0.060 \text{ V} - 0.059 \text{V} = 0.001 \text{V}$$

The standard deviation is

$$s_{V_{offset}} = \sqrt{\frac{(0.003 \text{ V})^2 + (0.002 \text{ V})^2 + (0.001 \text{ V})^2}{2}}$$
$$= \sqrt{\frac{0.000014 \text{ V}^2}{2}} = \sqrt{0.000007 \text{ V}^2} = 0.0026 \text{ V}.$$

Thus, the repeatability uncertainty is

$$u_{V_{ran}} = 0.0026 \text{ V}.$$

The mean voltage is

$$\overline{V} = V_0 + \overline{V}_{offset}$$
$$= (4.80 + 0.059) \text{ V}$$
$$= 4.859 \text{ V}$$

and the repeatability uncertainty in the mean voltage is

$$u_{\overline{V}_{ran}} = \frac{0.0026 \text{ V}}{\sqrt{3}} = \frac{0.0026 \text{ V}}{1.732} = 0.0015 \text{ V}.$$

The mean voltage is the reported output value in this analysis, so $u_{\overline{V}_{ran}}$ should be used for the combined uncertainty estimate. The estimated uncertainties for each parameter are listed in

Table 7-10.

Table 7-10. Estimated Uncertainties for Digital Multimeter Module Parameters

Parameter Name	Nominal or Mean Value	± Error Limits	Percent Conf.	Standard Uncertainty	Sensitivity Coefficient	Component Uncertainty
Amp_{Out}	4.80 V			0.0877 V	1	0.0877 V
DMM_{Acc}	0 V	± 0.0054 V	95	0.0027 V	1	0.0027 V
DMM_{res}	0 V	± 0.0005 V	100	0.0003 V	1	0.0003 V
V_{ran}	0.059 V			0.0015 V	1	0.0015 V

The average DMM output voltage is 4.859 V and the uncertainty in this value is computed by taking the root sum square of the standard uncertainties.

$$u_{DMM_{Out}} = \sqrt{(0.0877 \text{ V})^2 + (0.0027 \text{ V})^2 + (0.0003 \text{ V})^2 + (0.0015 \text{ V})^2}$$

$$= \sqrt{0.0077 \text{ V}^2} = 0.0878 \text{ V}.$$

The degrees of freedom for the DMM output uncertainty are computed using the Welch-Satterthwaite formula given in equation (7-18).

$$\nu_{DMM_{Out}} = \frac{u_{DMM_{Out}}^4}{\dfrac{u_{Amp_{Out}}^4}{\nu_{Amp_{Out}}} + \dfrac{u_{DMM_{Acc}}^4}{\nu_{DMM_{Acc}}} + \dfrac{u_{DMM_{res}}^4}{\nu_{DMM_{res}}} + \dfrac{u_{\bar{V}_{ran}}^4}{\nu_{\bar{V}_{ran}}}} \tag{7-18}$$

The degrees of freedom for error source uncertainties were assumed to be infinite, except for the uncertainty due to repeatability error, which has a degrees of freedom equal to 2. So, the degrees of freedom for the estimated uncertainty in the DMM output voltage is computed to be

$$\nu_{DMM} = \nu_{\bar{V}_{ran}} \times \frac{u_{DMM_{Out}}^4}{u_{\bar{V}_{ran}}^4} = 2 \times \left(\frac{0.0877 \text{ V}}{0.0015 \text{ V}}\right)^4 = 2 \times (58.5)^4 \cong \infty.$$

7.9 Compute System Output Uncertainty

In general, the system output uncertainty is equal to the output uncertainty for the final module. The associated degrees of freedom for the system output uncertainty are also equal to the degrees of freedom for the final module output uncertainty.

In the evaluation of the load cell system modules, it has been illustrated how the uncertainty in the output of one module propagates through to the next module in the series. For a 3 lb$_f$ input load or weight, the average system output, $\bar{V}$, and output uncertainty, $u_{\bar{V}}$, are 4.859 V and 0.097 V (or 97 mV), respectively.

Note: The load cell system analysis can be duplicated for other calibration weights. The resulting input weights, output voltages and uncertainties could then be used to create uncertainty statements for a range of values.

7.10   Report Analysis Results

The analysis results for the load cell measurement system are summarized in Table 7-11. As should be expected, the signal output uncertainty increases substantially as errors propagate through the amplifier module.

Table 7-11. Summary of Results for Load Cell System Analysis

Module Name	Module Input	Module Output	Standard Uncertainty	Degrees of Freedom
Load Cell	3 lb$_f$	9.60 mV	0.174 mV	∞
Amplifier	9.60 mV	4.80 V	87.7 mV	∞
Digital Multimeter	4.80 V	4.859 V	87.8 mV	∞

It is useful to take a closer look to determine how the uncertainties for each module contribute to the overall system output uncertainty. This can be accomplished by viewing the pareto chart for each module, shown in Figures 7-4 through 7-6.

The pareto chart for the load cell module shows that the excitation voltage and zero balance are the largest contributors to the load cell output uncertainty. Replacement of the TMO-2 excitation voltage with a precision voltage source could significantly reduce the load cell output uncertainty. Mitigation of the zero balance error, however, would most likely require a different load cell.

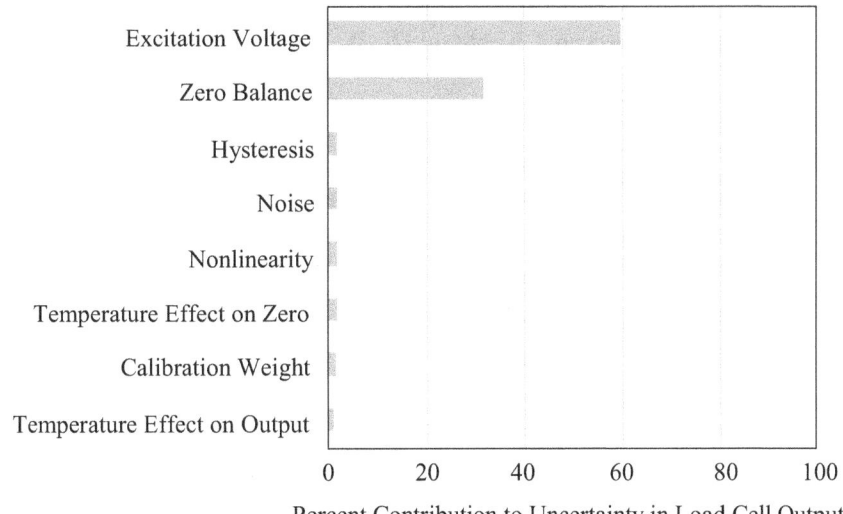

Percent Contribution to Uncertainty in Load Cell Output

Figure 7-4. Pareto Chart for Load Cell Module

Because the load cell output uncertainty is multiplied by the amplifier gain, it is the largest contributor to the amplifier output uncertainty, as shown in Figure 7-5. Errors due to amplifier balance stability and temperature coefficient also have some effect on the amplifier output uncertainty.

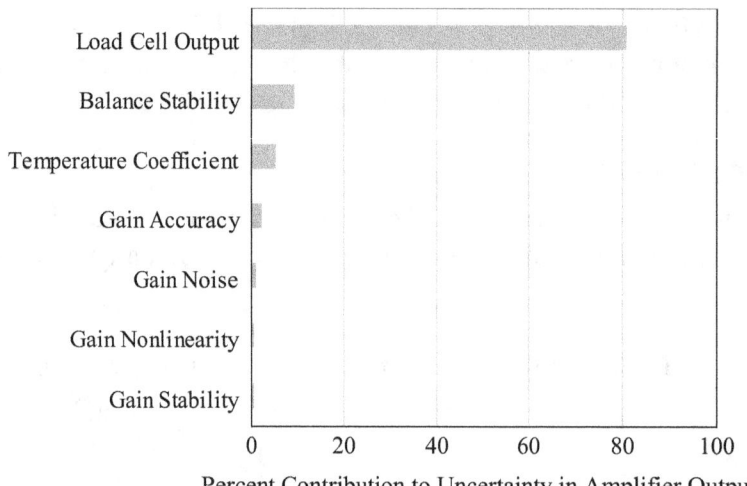

Percent Contribution to Uncertainty in Amplifier Output

Figure 7-5. Pareto Chart for Amplifier Module

As expected, the amplifier output uncertainty is the largest contributor to the digital multimeter output uncertainty. The accuracy of the digital multimeter also adds to the output uncertainty.

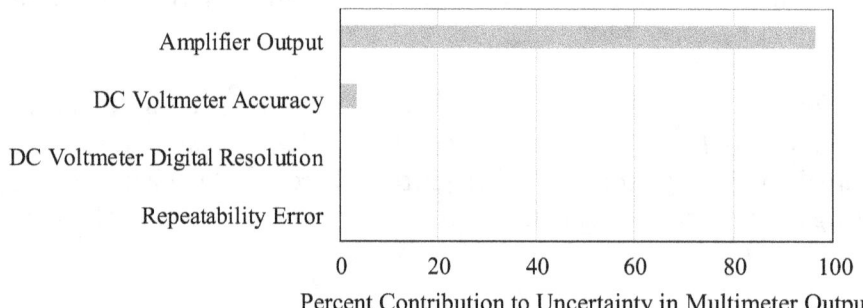

Percent Contribution to Uncertainty in Multimeter Output

Figure 7-6. Pareto Chart for Digital Multimeter Module

7.10.1 Confidence Limits

The system output uncertainty and degrees of freedom can be used to compute confidence limits that are expected to contain the system output voltage with some specified confidence level or probability, p. The confidence limits are expressed as

$$\bar{V} \pm t_{\alpha/2,\nu} u_{\bar{V}} \tag{7-19}$$

where the multiplier, $t_{\alpha/2\nu}$, is the t-statistic and $\alpha = 1 - p$.

For this analysis, let us assume that we want 95% confidence limits (i.e., $p = 0.95$). The corresponding t-statistic is $t_{0.025,\infty} = 1.96$ and the confidence limits are computed to be

$$4.859\text{ V} \pm 1.96 \times 0.0878\text{ V} \text{ or } 4.859\text{ V} \pm 0.172\text{ V}.$$

CHAPTER 8: UNCERTAINTY ANALYSIS FOR ALTERNATIVE CALIBRATION SCENARIOS

Calibrations are performed to obtain an estimate of the value or bias of selected unit-under-test (UUT) attributes.[54] In general, calibrations are not considered complete without statements of the uncertainty in these estimates. Developing these statements requires that all relevant sources of measurement error are identified and combined in a way that yields viable uncertainty estimates.

Unfortunately, confusion regarding which error sources should be included and how they should be combined often exists for calibration processes. Much of this confusion can be eliminated by an examination of the objective of each UUT attribute calibration and a consideration of the corresponding measurement configuration or "scenario."

In this chapter, the calibration of a UUT attribute is examined within the context of four scenarios.

1. The measurement reference (referred herein as the MTE) measures the value of the UUT attribute.
2. The UUT measures the value of the MTE attribute.
3. The UUT and MTE attribute values are measured with a comparator.
4. The UUT and MTE both measure the value of an attribute of a common artifact.

Each scenario yields an observed value, referred to as a "measurement result" or "calibration result" and a description of measurement process errors that accompany this result. This information is summarized and then employed to obtain an uncertainty estimate in the calibration result. Examples are given to illustrate concepts and procedures.

8.1   Calibration Scenarios Overview

The four calibration scenarios listed above are described in detail in the following sections. The descriptions provide guidelines for developing uncertainty estimates relevant to each scenario. The structure and content of each description is intended to provide a basis for developing whatever mathematical customization is needed for specific measurement situations.

In each scenario, we have a measurement denoted δ. The general measurement equation is

$$\delta = e_{UUT,b} + \varepsilon_{cal} \tag{8-1}$$

where $e_{UUT,b}$ is the true UUT attribute bias and ε_{cal} is the calibration error. Applying the variance operator to equation (8-1), the uncertainty in δ is

$$
\begin{aligned}
u_{cal} = \sqrt{\mathrm{var}(\delta)} &= \sqrt{\mathrm{var}(e_{UUT,b} + \varepsilon_{cal})} \\
&= \sqrt{\mathrm{var}(\varepsilon_{cal})}.
\end{aligned} \tag{8-2}
$$

---

[54] An attribute is a measurable characteristic, feature or aspect of an object or substance.

8.1.1 Special Notation

The notation used in this chapter differs slightly from that used in other chapters and appendices in this document. The subscripts and variables designators used in this chapter are summarized in Table 8-1.

Table 8-1. Calibration Scenarios Notation

Notation	Description
e	an individual measurement process error, such as repeatability, resolution error, etc.
ε	combined errors comprised of individual measurement process errors
m	measurement
b	bias
cal	calibration
$true$	true value
n	nominal value
x	quantities relating to the UUT
y	quantities relating to the MTE

This special notation is intended to provide a means of distinguishing between individual measurement process errors and combined errors. For example, measurement error is represented by the quantity ε_m, the error in a calibration result is represented by ε_{cal} and the bias in the UUT attribute is represented by the quantity $e_{UUT,b}$.

8.1.2 Measurement Error Sources

Measurement process errors encountered in a given calibration scenario typically include:[55]

$e_{MTE,b}$ = bias in the measurement reference or MTE

e_{rep} = repeatability or random error

e_{res} = resolution error

e_{op} = operator bias

e_{other} = other measurement error, such as that due to environmental corrections, ancillary equipment variations, response to adjustments, etc.

As discussed in Chapter 2, the sum of the errors encountered during the measurement process can be expressed as

$$\varepsilon_m = e_{MTE,b} + e_{rep} + e_{res} + e_{op} + e_{other} \qquad (8\text{-}3)$$

where equation (8-3) is the measurement error model.

---

[55] Descriptions of these measurement process errors are given in Chapter 3.

8.1.3 Calibration Error and Measurement Error

As previously discussed, the result of a calibration is taken to be an estimation of the true UUT attribute bias, $e_{UUT,b}$. The error in the calibration result is represented by the quantity ε_{cal}. In all four calibration scenarios, the uncertainty in the estimation of $e_{UUT,b}$ is computed as the uncertainty in ε_{cal}.

For some calibration scenarios, ε_{cal} is synonymous with the measurement error ε_m. However, in other scenarios, ε_{cal} and ε_m may not have equivalent sign or magnitude.

8.1.4 UUT Attribute Bias

For calibrations, it is implicitly assumed that the UUT attribute of interest is assigned some design or "nominal" value x_n. The true value of the UUT attribute, x_{true}, is the nominal value plus the UUT attribute bias.

$$x_{true} = x_n + e_{UUT,b} \tag{8-4}$$

The difference between the UUT attribute's true value, x_{true}, and the nominal value x_n is the UUT attribute's bias $e_{UUT,b}$.

$$e_{UUT,b} = x_{true} - x_n \tag{8-5}$$

> **Note**: Equation (8-4) does not represent the basic measurement equation $x_n = x_{true} + \varepsilon_m$. Rather, it is a statement of the relationship between the UUT attribute's true value, its stated nominal value and its bias. In this context, the relationship between measurement error and the UUT attribute bias is $\varepsilon_m = - e_{UUT,b}$.

In some cases, the UUT is a passive item, such as a gage block or weight, whose attribute of interest is a simple characteristic like length or mass. In other cases, the UUT is an active device such as a voltmeter or tape measure, whose attribute consists of a reading or other output like voltage or measured length. In the former case, the concepts of true value and nominal value are straightforward. In the latter case, some comment is needed.

As stated earlier, the result of a calibration is considered to be an estimate of the quantity $e_{UUT,b}$. From equation (8-4), if the UUT attribute has a nominal value x_n, estimating x_{true} is equivalent to estimating $e_{UUT,b}$. Additionally, $e_{UUT,b}$ is an "inherent" property of the UUT attribute, independent of its resolution, repeatability or other characteristic dependent on its application or usage environment.

Accordingly, if the UUT's nominal value consists of a measured reading or other actively displayed output, the UUT bias must be taken to be the difference between the true value of the quantity being measured and the value internally sensed by the UUT, with appropriate environmental or other adjustments applied to correct this value to reference (calibration) conditions.

For example, suppose the UUT is a steel yardstick whose length is a random variable following a probability distribution with a standard deviation arising from variations in the manufacturing process. The UUT is used under specified nominal environmental conditions where repeatability, resolution error, operator bias and other error sources may come into play. In this case, the bias of the yardstick is systematically present, regardless of whatever chance relationship may exist between the length of the measured object, the closest observed "tick mark," the temperature of the measuring environment, the perspective of the operator, and so on.

8.1.5   MTE Attribute Bias

The value of the MTE attribute, which the value of the UUT attribute is compared against, has an inherent deviation $e_{MTE,b}$ from its nominal attribute value y_n or value stated in a calibration certificate or other document. The true value of the MTE attribute y_{true} is the nominal value plus the MTE attribute bias.

$$y_{true} = y_n + e_{MTE,b} \qquad (8\text{-}6)$$

> **Note**: As with Equation (8-4), Equation (8-6) does not represent the basic measurement equation $x_n = x_{true} + \varepsilon_m$.

As with the UUT, the MTE may be a passive item, such as a gage block or weight or an active device, such as a voltmeter or tape measure. In either case, it is important to bear in mind that $e_{MTE,b}$ is an inherent property of the MTE attribute, exclusive of other errors such as MTE resolution or the repeatability of the measurement process. The value of the MTE attribute may vary with environmental deviations, but it can usually be adjusted or corrected to some reference set of conditions.

8.2   Scenario 1: The MTE Measures the UUT Attribute Value

In this calibration scenario, the UUT is a passive device whose attribute provides no reading or other metered output. Its output may consist of a generated value, as in the case of a voltage reference, or a fixed value, as in the case of a gage block.[56] The measurement equation is

$$y = x_{true} + \varepsilon_m \qquad (8\text{-}7)$$

where y is the measurement result obtained with the MTE, x_{true} is the true value of the UUT attribute and ε_m is the measurement error.

Substituting equation (8-4) into equation (8-7), the measurement equation can be written as

$$y = x_n + e_{UUT,b} + \varepsilon_m. \qquad (8\text{-}8)$$

The difference $y - x_n$ is a measurement of the UUT attribute bias $e_{UUT,b}$. This quantity is denoted by the variable δ and defined as

---

[56] Cases where the MTE measures the value of a metered or other UUT attribute exhibiting a displayed value are covered later as special instances of Scenario 4.

$$\delta = y - x_n$$
$$= e_{UUT,b} + \varepsilon_m \tag{8-9}$$
$$= e_{UUT,b} + \varepsilon_{cal}$$

where

$$\varepsilon_{cal} = \varepsilon_m = e_{MTE,b} + e_{rep} + e_{res} + e_{op} + e_{other}. \tag{8-10}$$

Since the UUT is a passive device, resolution error and operator bias arise exclusively from the use of the MTE. In addition, the uncertainty due to repeatability is estimated from a random sample of measurements taken with the MTE. However, variations in UUT attribute value may contribute to this estimate. Random variations in UUT attribute value and random variations due to other causes are not separable from random variations due to the MTE.[57] Consequently, e_{rep} must be taken to represent a "measurement process error" rather than an error attributable to any specific influence.

Given these considerations, the error sources e_{rep}, e_{res} and e_{op} in equation (8-10) are

$$e_{rep} = e_{MTE,rep}$$
$$e_{res} = e_{MTE,res} \tag{8-11}$$
$$e_{op} = e_{MTE,op}$$

where $e_{MTE,rep}$ represents the repeatability of the measurement process. The "MTE" part of the subscript indicates that the uncertainty in the error will be estimated from a sample of measurements taken by the MTE.

From equations (8-10) and (8-11), the error in the calibration result δ is

$$\varepsilon_{cal} = e_{MTE,b} + e_{MTE,rep} + e_{MTE,res} + e_{MTE,op} + e_{other} \tag{8-12}$$

and the uncertainty in δ is

$$u_{cal} = \sqrt{\mathrm{var}(\varepsilon_{cal})}$$
$$= \sqrt{\mathrm{var}(e_{MTE,b}) + \mathrm{var}(e_{MTE,rep}) + \mathrm{var}(e_{MTE,res}) + \mathrm{var}(e_{MTE,op}) + \mathrm{var}(e_{other})} \tag{8-13}$$
$$= \sqrt{u_{MTE,b}^2 + u_{MTE,rep}^2 + u_{MTE,res}^2 + u_{MTE,op}^2 + u_{other}^2}.$$

The error source e_{other} may arise from corrections ensuing from environmental factors, such as thermal expansion. In this case, it may be necessary to correct measured values to those that would be attained at some reference temperature, such as 20 °C.

For example, let the UUT attribute be gage block length and the MTE attribute be the reading obtained with a super micrometer. If $\delta_{UUT,env}$ and $\delta_{MTE,env}$ represent thermal expansion

---

[57] As stated in Section 2.3, random variations in a measured quantity are not separable from random variations due other error sources.

corrections to the UUT and MTE attributes, respectively, then the mean value of the measurement sample would be corrected by an amount equal to[58]

$$\delta_{env} = \delta_{MTE,env} - \delta_{UUT,env} \qquad (8\text{-}14)$$

and the error in the environmental correction δ_{env} would be written

$$e_{other} = e_{env} = e_{MTE,env} - e_{UUT,env}. \qquad (8\text{-}15)$$

The error in the corrected calibration result $\delta_{corr} = \delta - \delta_{env}$ is

$$\varepsilon_{cal} = e_{MTE,b} + e_{MTE,rep} + e_{MTE,res} + e_{MTE,op} + e_{MTE,env} - e_{UUT,env} \qquad (8\text{-}16)$$

and the uncertainty in δ_{corr} is

$$
\begin{aligned}
u_{cal} &= \sqrt{\operatorname{var}(\varepsilon_{cal})} \\
&= \sqrt{\operatorname{var}(e_{MTE,b}) + \operatorname{var}(e_{MTE,rep}) + \operatorname{var}(e_{MTE,res}) + \operatorname{var}(e_{MTE,op}) + \operatorname{var}(e_{MTE,env} - e_{UUT,env})} \\
&= \sqrt{u_{MTE,b}^2 + u_{MTE,rep}^2 + u_{MTE,res}^2 + u_{MTE,op}^2 + u_{MTE,env}^2 + u_{UUT,env}^2 - 2\rho_{env} u_{MTE,env} u_{UUT,env}}
\end{aligned}
\qquad (8\text{-}17)
$$

where the correlation coefficient ρ_{env} accounts for any correlation between $e_{MTE,env}$ and $e_{UUT,env}$. The correlation coefficient can range in value from -1 to $+1$. If the same temperature measurement device (e.g., thermometer) is used to make both the UUT and MTE corrections, then

$$\rho_{env} = 1 \qquad (8\text{-}18)$$

and equation (8-17) can be rewritten as

$$u_{cal} = \sqrt{u_{MTE,b}^2 + u_{MTE,rep}^2 + u_{MTE,res}^2 + u_{MTE,op}^2 + u_{MTE,env}^2 + u_{UUT,env}^2 - 2u_{MTE,env} u_{UUT,env}}. \qquad (8\text{-}19)$$

8.3   Scenario 2 : The UUT Measures the MTE Attribute Value

In this scenario, the MTE is a passive device whose reference attribute provides no reading or other metered output. Its output may consist of a generated value, as in the case of a voltage reference, or a fixed value, as in the case of a gage block.[59] The measurement equation is

$$x = y_{true} + \varepsilon_m \qquad (8\text{-}20)$$

---

[58] The form of this expression arises from the fact that thermal expansion of the gage block results in an inflated gage block length, while thermal expansion of the micrometer results in applying additional thimble adjustments to narrow the gap between the anvil and the spindle, resulting in a deflated measurement reading.

[59] Cases where the UUT measures the value of a metered or other MTE attribute exhibiting a displayed value are covered later as special instances of Scenario 4.

where x is the value measured by the UUT, y_{true} is the true value of the MTE attribute being measured and ε_m is the measurement error. Substituting equation (8-6) into equation (8-20), the measurement equation can be written as

$$x = y_n + e_{MTE,b} + \varepsilon_m. \tag{8-21}$$

The difference $x - y_n$ is a measurement of the UUT attribute bias $e_{UUT,b}$. This calibration result is denoted by the variable δ and defined by

$$\delta = x - y_n = e_{MTE,b} + \varepsilon_m \tag{8-22}$$

For this scenario, the measurement error model is

$$\varepsilon_m = e_{UUT,b} + e_{rep} + e_{res} + e_{op} + e_{other} \tag{8-23}$$

where $e_{UUT,b}$ is the UUT attribute bias. In this scenario, the MTE is a passive device. Therefore, resolution error and operator bias arise exclusively from the use of the UUT. In addition, the uncertainty due to repeatability is estimated from a random sample of measurements taken with the UUT. Consequently, the error sources e_{rep}, e_{res} and e_{op} in equation (8-23) are

$$\begin{aligned} e_{rep} &= e_{UUT,rep} \\ e_{res} &= e_{UUT,res} \\ e_{op} &= e_{UUT,op} \,. \end{aligned} \tag{8-24}$$

The "UUT" part of the subscript indicates that the uncertainty in the error will be estimated from a sample of measurements taken by the UUT. The error source e_{other} may need to include mixed contributions as described in Scenario 1.

Substituting equations (8-23) and (8-24) into equation (8-22) and rearranging gives

$$\delta = e_{UUT,b} + e_{MTE,b} + e_{UUT,rep} + e_{UUT,res} + e_{UUT,op} + e_{other} \tag{8-25}$$

As in scenario 1, equation (8-25) provides an expression that is separable into a measurement δ of the UUT attribute bias, $e_{UUT,b}$, and a calibration error, ε_{cal}, given by

$$\delta = e_{UUT,b} + \varepsilon_{cal} \tag{8-26}$$

where

$$\varepsilon_{cal} = e_{MTE,b} + e_{UUT,rep} + e_{UUT,res} + e_{UUT,op} + e_{other} \,. \tag{8-27}$$

The uncertainty in δ, and thus, $e_{UUT,b}$ is

$$u_{cal} = \sqrt{\mathrm{var}(\varepsilon_{cal})}$$
$$= \sqrt{\mathrm{var}(\varepsilon_{MTE,b}) + \mathrm{var}(\varepsilon_{UUT,rep}) + \mathrm{var}(\varepsilon_{UUT,res}) + \mathrm{var}(\varepsilon_{UUT,op}) + \mathrm{var}(\varepsilon_{other})} \qquad (8\text{-}28)$$
$$= \sqrt{u_{MTE,b}^2 + u_{UUT,rep}^2 + u_{UUT,res}^2 + u_{UUT,op}^2 + u_{other}^2}.$$

8.4   Scenario 3: The MTE and UUT Attribute Values are Compared

In this scenario, a device called a "comparator" is used to measure or compare UUT and MTE attribute values.[60] In keeping with the basic notation, the indicated value of the UUT attribute x is expressed as

$$x = x_{true} + \varepsilon_{UUT,m} \qquad (8\text{-}29)$$

and the indicated value of the MTE attribute y is expressed as

$$y = y_{true} + \varepsilon_{MTE,m} \qquad (8\text{-}30)$$

where $\varepsilon_{UUT,m}$ is the measurement error involved in the use of the comparator to measure the UUT attribute value and $\varepsilon_{MTE,m}$ is the measurement error involved in the use of the comparator to measure the MTE attribute value.

As discussed in Sections 8.1.4 and 8.1.5, for calibrations, the UUT attribute and MTE attribute are assigned some design or "nominal" values x_n and y_n, respectively. Substituting equation (8-4) into equation (8-29) gives

$$x = x_n + e_{UUT,b} + \varepsilon_{UUT,m}. \qquad (8\text{-}31)$$

Similarly, substituting equation (8-6) into equation (8-30) gives

$$y = y_n + e_{MTE,b} + \varepsilon_{MTE,m}. \qquad (8\text{-}32)$$

The result of the comparison is a measured deviation δ, which is expressed as

$$\begin{aligned} \delta &= x - y \\ &= x_n - y_n + e_{UUT,b} - e_{MTE,b} + (\varepsilon_{UUT,m} - \varepsilon_{MTE,m}). \end{aligned} \qquad (8\text{-}33)$$

In most calibrations involving comparators, $x_n = y_n$ and equation (8-33) becomes [61]

$$\delta = e_{UUT,b} - e_{MTE,b} + (\varepsilon_{UUT,m} - \varepsilon_{MTE,m}). \qquad (8\text{-}34)$$

As with the previous scenarios, equation (8-34) provides an expression that is separable into a measurement δ of the UUT attribute bias, $e_{UUT,b}$, and a calibration error, ε_{cal}, given by

---

[60] The MTE and UUT attributes may be measured sequentially or simultaneously, depending on the comparator device.
[61] To accommodate cases where $y_n \neq x_n$, $\delta = (x - x_n) - (y - y_n)$. For example, consider a case where the MTE is a 2 cm gage block and the UUT is a 1 cm gage block. Suppose that the comparator readings for the MTE and UUT are 2.10 cm and 0.99 cm, respectively. Then, $\delta = (0.99 - 1.0) - (2.10 - 2.0) = -0.110$ cm. The corrected value for the UUT attribute is $x_c = x + \delta = 1.0$ cm $+ (-0.110)$ cm $= 0.89$ cm.

$$\delta = e_{UUT,b} + \varepsilon_{cal} \tag{8-35}$$

where

$$\varepsilon_{cal} = \left(\varepsilon_{UUT,m} - \varepsilon_{MTE,m}\right) - e_{MTE,b}. \tag{8-36}$$

The measurement error model for $\varepsilon_{MTE,m}$ is

$$\varepsilon_{MTE,m} = e_{c,b} + e_{MTE,rep} + e_{MTE,res} + e_{MTE,op} + e_{MTE,other} \tag{8-37}$$

where $e_{c,b}$ represent the bias of the comparator. Similarly, the measurement error model for $\varepsilon_{UUT,m}$ is

$$\varepsilon_{UUT,m} = e_{c,b} + e_{UUT,rep} + e_{UUT,res} + e_{UUT,op} + e_{UUT,other}. \tag{8-38}$$

Substituting equations (8-37) and (8-38) into equation (8-36), ε_{cal} is

$$
\begin{aligned}
\varepsilon_{cal} = &\left(e_{UUT,rep} - e_{MTE,rep}\right) + \left(e_{UUT,res} - e_{MTE,res}\right) + \left(e_{UUT,op} - e_{MTE,op}\right) \\
&+ \left(e_{UUT,other} - e_{MTE,other}\right) - e_{MTE,b}.
\end{aligned}
\tag{8-39}
$$

The uncertainty in δ is

$$
\begin{aligned}
u_{cal} &= \sqrt{\mathrm{var}(\varepsilon_{cal})} \\
&= \sqrt{\begin{aligned}&\mathrm{var}(e_{UUT,rep} - e_{MTE,rep}) + \mathrm{var}(e_{UUT,res} - e_{MTE,res}) + \mathrm{var}(e_{UUT,op} - e_{MTE,op}) \\ &+ \mathrm{var}(e_{UUT,other} - e_{MTE,other}) + \mathrm{var}(-e_{MTE,b}).\end{aligned}}
\end{aligned}
\tag{8-40}
$$

Accounting for possible correlations between $e_{UUT,op}$ and $e_{MTE,op}$ and between $e_{UUT,other}$ and $e_{MTE,other}$, the uncertainty in δ can be expressed as

$$
u_{cal} = \sqrt{\begin{aligned}&u_{MTE,rep}^2 + u_{UUT,rep}^2 + u_{MTE,res}^2 + u_{UUT,res}^2 + u_{MTE,op}^2 + u_{UUT,op}^2 \\ &-2\rho_{op}u_{MTE,op}u_{UUT,op} + u_{MTE,other}^2 + u_{UUT,other}^2 - 2\rho_{other}u_{MTE,other}u_{UUT,other} + u_{MTE,b}^2\end{aligned}}. \tag{8-41}
$$

8.5   Scenario 4: The MTE and UUT Measure a Common Artifact

In this scenario, both the MTE and UUT measure the attribute value of a common artifact. The measurements by the MTE and UUT are made and recorded separately. An example of this scenario is the calibration of a thermometer (UUT) using a temperature reference (MTE), where both the UUT and MTE are placed in an oven.

Denoting the true value of the artifact as T, the UUT measurement equation is

$$x = T + \varepsilon_{UUT,m} \tag{8-42}$$

where $\varepsilon_{UUT,m}$ is the measurement process error for the UUT measurement of the artifact's value.

Similarly, the MTE measurement equation is

$$y = T + \varepsilon_{MTE,m} \tag{8-43}$$

where $\varepsilon_{MTE,m}$ is the measurement process error for the MTE measurement of the artifact's value.

The difference between the measurement results δ is expressed as

$$\begin{aligned} \delta &= x - y \\ &= \varepsilon_{UUT,m} - \varepsilon_{MTE,m}. \end{aligned} \tag{8-44}$$

The measurement error model for $\varepsilon_{UUT,m}$ is

$$\varepsilon_{UUT,m} = e_{UUT,b} + e_{UUT,rep} + e_{UUT,res} + e_{UUT,op} + e_{UUT,other} \tag{8-45}$$

and the measurement error model for $\varepsilon_{MTE,m}$ is

$$\varepsilon_{MTE,m} = e_{MTE,b} + e_{MTE,rep} + e_{MTE,res} + e_{MTE,op} + e_{MTE,other}. \tag{8-46}$$

Substituting equations (8-45) and (8-46) into equation (8-44), provides an expression that is separable into a measurement δ of the UUT attribute bias, $e_{UUT,b}$, and a calibration error, ε_{cal}, given by

$$\delta = e_{UUT,b} + \varepsilon_{cal} \tag{8-47}$$

where

$$\begin{aligned} \varepsilon_{cal} &= -e_{MTE,b} + (e_{UUT,rep} - e_{MTE,rep}) + (e_{UUT,res} - e_{MTE,res}) \\ &\quad + (e_{UUT,op} - e_{MTE,op}) + (e_{UUT,other} - e_{MTE,other}). \end{aligned} \tag{8-48}$$

As in scenario 3, the uncertainty in δ is

$$\begin{aligned} u_{cal} &= \sqrt{\operatorname{var}(\varepsilon_{cal})} \\ &= \sqrt{\begin{array}{l} \operatorname{var}(-e_{MTE,b}) + \operatorname{var}(e_{UUT,rep} - e_{MTE,rep}) + \operatorname{var}(e_{UUT,res} - e_{MTE,res}) \\ + \operatorname{var}(e_{UUT,op} - e_{MTE,op}) + \operatorname{var}(e_{UUT,other} - e_{MTE,other}). \end{array}} \end{aligned} \tag{8-49}$$

Accounting for possible correlations between $e_{UUT,op}$ and $e_{MTE,op}$ and between $e_{UUT,other}$ and $e_{MTE,other}$, the uncertainty in δ can be expressed as

$$u_{cal} = \sqrt{\begin{array}{l} u_{MTE,b}^2 + u_{MTE,rep}^2 + u_{UUT,rep}^2 + u_{MTE,res}^2 + u_{UUT,res}^2 + u_{MTE,op}^2 + u_{UUT,op}^2 \\ -2\rho_{op} u_{MTE,op} u_{UUT,op} + u_{MTE,other}^2 + u_{UUT,other}^2 - 2\rho_{other} u_{MTE,other} u_{UUT,other} \end{array}}. \tag{8-50}$$

8.5.1 Special Cases for Scenario 4

There are two special cases of Scenario 4 that may be thought of as variations of Scenarios 1 and 2. Both cases are accommodated by the Scenario 4 definitions and expressions previously developed.

Case 1: The MTE measures the UUT and both the MTE and UUT provide a metered or other displayed output.

In this case, the common artifact is the UUT attribute, consisting of a "stimulus" embedded in the UUT. An example would be a UUT voltage source whose output is indicated by a digital display and is measured using an MTE voltmeter.

Case 2: The UUT measures the MTE and both the MTE and UUT provide a metered or other displayed output.

In this case, the common artifact is the MTE attribute, consisting of a "stimulus" embedded in the MTE. An example would be an MTE voltage source whose output is indicated by a digital display and is measured using a UUT voltmeter.

8.6   Uncertainty Analysis Examples

Four scenarios have been discussed that yield expressions for calibration uncertainty. In all scenarios, the calibration result is expressed as

$$\delta = e_{UUT,b} + \varepsilon_{cal}$$

and the calibration uncertainty is

$$u_{cal} = \sqrt{\text{var}(\varepsilon_{cal})} \, .$$

Uncertainty analysis examples for the four calibration scenarios are provided in the following subsections.

8.6.1   Scenario 1: The MTE Measures the UUT Attribute Value

In this scenario, the measurement result is $\delta = y - x_n$ and ε_{cal} is expressed in equation (8-12). The example for this scenario consists of calibrating a 30 gm mass with a precision balance. The uncertainty in the local gravity is considered to be negligible in this measurement process. Multiple measurements of the UUT mass are taken and the sample statistics are computed to be

Sample Mean	= 30.000047 gm
Standard Deviation	= 1.15×10^{-5} gm
Uncertainty in the Mean	= 6.64×10^{-6} gm
Sample Size	= 3

The measurement result is $\bar{\delta} = (30.000047 - 30)$ gm $= 4.7 \times 10^{-5}$ gm. However, the measurements are not taken in a vacuum, so the buoyancy of displaced air can introduce measurement error. The balance is calibrated with calibration weights with a density of $\rho_{wt} = 8.0$ gm/cm^3. The air buoyancy correction is

$$\bar{y}_{corr} = \bar{y} \times \frac{\left(1 - \rho_{air} / \rho_{wt}\right)}{\left(1 - \rho_{air} / \rho_{UUT}\right)}$$

where $\bar{y}$ is the sample mean, ρ_{air} is the local air density and ρ_{UUT} is the density of the UUT mass. For this analysis, we will assume that $\rho_{air} = 1.2 \times 10^{-3}$ gm/cm^3 and $\rho_{UUT} = 8.4$ gm/cm^3. The corrected sample mean is computed to be

$$\bar{y}_{corr} = 30.000047 \text{ gm} \times \frac{\left(1 - 0.0012 / 8.0\right)}{\left(1 - 0.0012 / 8.4\right)}$$

$$= 30.000047 \text{ gm} \times \frac{\left(1 - 0.00015\right)}{\left(1 - 0.00014\right)}$$

$$= 30.000047 \text{ gm} \times \frac{0.99985}{0.99986} = 30.000047 \text{ gm} \times 0.99999$$

$$= 29.99975 \text{ gm}$$

and the corrected calibration result is $\bar{\delta}_{corr} = (29.99975 - 30) \text{ gm} = -2.5 \times 10^{-4} \text{ gm}$.

In the mass calibration scenario, the following measurement process errors must be considered:

- Bias of the precision balance, $e_{MTE,b}$.
- Repeatability, $e_{MTE,rep}$.
- Error due to the digital resolution of the balance, $e_{MTE,res}$.
- Environmental factors error resulting from the buoyancy correction, e_{env}.

The error in $\bar{\delta}_{corr}$ is

$$\varepsilon_{cal} = e_{MTE,b} + e_{MTE,rep} + e_{MTE,res} + e_{env}$$

where

$$e_{env} = e_{UUT,env} = c_1 e_{\rho_{air}} + c_2 e_{\rho_{UUT}}$$

and $e_{\rho_{air}}$ and $e_{\rho_{UUT}}$ are the errors in the air and UUT densities, respectively. The coefficients c_1 and c_2 are sensitivity coefficients that determine the relative contribution of the errors $e_{\rho_{air}}$ and $e_{\rho_{UUT}}$ to the total error e_{env}. The uncertainty in $\bar{\delta}_{corr}$ is

$$\begin{aligned} u_{cal} &= \sqrt{\mathrm{var}(\varepsilon_{cal})} \\ &= \sqrt{\mathrm{var}(e_{MTE,b}) + \mathrm{var}(e_{MTE,rep}) + \mathrm{var}(e_{MTE,res}) + \mathrm{var}(c_1 e_{\rho_{air}} + c_2 e_{\rho_{UUT}})} \\ &= \sqrt{u_{MTE,b}^2 + u_{MTE,rep}^2 + u_{MTE,res}^2 + c_1^2 u_{\rho_{air}}^2 + c_2^2 u_{\rho_{UUT}}^2 + 2 c_1 c_2 \rho_{env} u_{\rho_{air}} u_{\rho_{UUT}}} . \end{aligned}$$

The correlation coefficient ρ_{env} accounts for any correlation between $e_{\rho_{air}}$ and $e_{\rho_{UUT}}$. The correlation coefficient can range in value from -1 to $+1$. In this analysis, the error in the air density is considered to be uncorrelated to the error in the density of the UUT mass. Therefore, $\rho_{env} = 0$ and the uncertainty u_{cal} can be expressed as

$$u_{cal} = \sqrt{u_{MTE,b}^2 + u_{MTE,rep}^2 + u_{MTE,res}^2 + c_1^2 u_{\rho_{air}}^2 + c_2^2 u_{\rho_{UUT}}^2}$$

$$= \sqrt{u_{MTE,b}^2 + u_{MTE,rep}^2 + u_{MTE,res}^2 + \left(c_1 u_{\rho_{air}}\right)^2 + \left(c_2 u_{\rho_{UUT}}\right)^2}$$

The sensitivity coefficients are computed to be[62]

$$c_1 = \frac{\partial \bar{y}_{corr}}{\partial \rho_{air}} = \frac{\partial}{\partial \rho_{air}}\left[\bar{y} \times \frac{1 - \rho_{air}/\rho_{wt}}{\left(1 - \rho_{air}/\rho_{UUT}\right)} \right] = \bar{y} \times \frac{\partial}{\partial \rho_{air}}\left[\frac{1 - \rho_{air}/\rho_{wt}}{\left(1 - \rho_{air}/\rho_{UUT}\right)} \right]$$

$$= \frac{\bar{y}}{1 - \rho_{air}/\rho_{UUT}} \times \left[\frac{1 - \rho_{air}/\rho_{wt}}{\rho_{UUT}\left(1 - \rho_{air}/\rho_{UUT}\right)} - \frac{1}{\rho_{wt}} \right]$$

$$= \frac{30.000047\,\text{gm}}{1 - 0.0012/8.4} \times \left[\frac{1 - 0.0012/8.0}{8.4\,\text{gm/cm}^3 \times \left(1 - 0.0012/8.4\right)} - \frac{1}{8.0\,\text{gm/cm}^3} \right]$$

$$= \frac{30.000047\,\text{gm}}{0.99986} \times \left[0.119\,\text{cm}^3/\text{gm} - 0.125\,\text{cm}^3/\text{gm} \right] = -0.18\,\text{cm}^3$$

$$c_2 = \frac{\partial \bar{y}_{corr}}{\partial \rho_{UUT}} = \frac{\partial}{\partial \rho_{UUT}}\left[\bar{y} \times \frac{1 - \rho_{air}/\rho_{wt}}{\left(1 - \rho_{air}/\rho_{UUT}\right)} \right] = \bar{y} \times \frac{\partial}{\partial \rho_{UUT}}\left[\frac{1 - \rho_{air}/\rho_{wt}}{\left(1 - \rho_{air}/\rho_{UUT}\right)} \right]$$

$$= -\bar{y} \times \frac{\rho_{air}}{\rho_{UUT}^2} \times \frac{1 - \rho_{air}/\rho_{wt}}{\left(1 - \rho_{air}/\rho_{UUT}\right)^2}$$

$$= -30.000047\,\text{gm} \times \frac{0.0012}{\left(8.4\right)^2}\,\text{cm}^3/\text{gm} \times \frac{1 - 0.0012/8.0}{\left(1 - 0.0012/8.4\right)^2}$$

$$= -30.000047\,\text{gm} \times 0.000017\,\text{cm}^3/\text{gm} \times \frac{0.99985}{0.99971} = -5.1 \times 10^{-4}\,\text{cm}^3$$

The distributions, limits, confidence levels and standard uncertainties for each error source are summarized in Table 8-2.

---

[62] Guidance on the development of multivariate error models is provided in Chapter 6.

Table 8-2. Summary of Scenario 1 Uncertainty Estimates

Error	Error Limits	Confidence Level (%)	Error Distribution	Deg. of Freedom	Analysis Type	Standard Uncertainty
$e_{MTE,b}$	± 0.12 gm	95.00	Normal	Infinite	B	6.12×10^{-2} gm
e_{rep}			Student's t	2	A	6.64×10^{-6} gm
e_{res}	± 0.005 gm	100.00	Uniform	Infinite	B	2.9×10^{-3} gm
$e_{\rho_{air}}$	$\pm 3.6 \times 10^{-5}$ gm/cm^3	95.00	Normal	Infinite	B	1.84×10^{-5} gm/cm^3
$e_{\rho_{UUT}}$	± 0.15 gm/cm^3	95.00	Normal	Infinite	B	0.077 gm/cm^3

Using the data in Table 8-2, the uncertainty in $\bar{\delta}_{corr}$ is computed to be

$$u_{cal} = \sqrt{\begin{array}{l}\left(6.12\times10^{-2}\right)^2 + \left(6.64\times10^{-6}\right)^2 + \left(2.9\times10^{-3}\right)^2 + \left(-0.18\times1.84\times10^{-5}\right)^2 \\ + \left(-5.1\times10^{-4}\times0.077\right)^2\end{array}} \text{ gm}$$

$$= \sqrt{3.75\times10^{-3} + 4.41\times10^{-11} + 8.41\times10^{-6} + 1.1\times10^{-11} + 1.54\times10^{-9}} \text{ gm}$$

$$= \sqrt{3.75\times10^{-3}} \text{ gm} = 6.12\times10^{-2} \text{ gm}.$$

The effective degrees of freedom ν_{eff} for u_{cal} can be estimated using the Welch-Satterthwaite formula.

$$\nu_{eff} = \frac{u_{cal}^4}{\dfrac{u_{MTE,b}^4}{\nu_{MTE,b}} + \dfrac{u_{MTE,rep}^4}{\nu_{MTE,rep}} + \dfrac{u_{MTE,res}^4}{\nu_{MTE,res}} + \dfrac{c_1^4 u_{\rho_{air}}^4}{\nu_{\rho_{air}}} + \dfrac{c_2^4 u_{\rho_{UUT}}^4}{\nu_{\rho_{UUT}}}}$$

$$= \frac{u_{cal}^4}{\dfrac{u_{MTE,b}^4}{\infty} + \dfrac{u_{MTE,rep}^4}{2} + \dfrac{u_{MTE,res}^4}{\infty} + \dfrac{c_1^4 u_{\rho_{air}}^4}{\infty} + \dfrac{c_2^4 u_{\rho_{UUT}}^4}{\infty}}$$

$$= 2 \times \frac{u_{cal}^4}{u_{MTE,rep}^4}$$

Therefore, the degrees of freedom are computed to be

$$\nu_{eff} = 2 \times \frac{\left(6.12\times10^{-2}\right)^4}{\left(6.64\times10^{-6}\right)^4} = 2 \times \left(0.92\times10^4\right)^4 = 2 \times 9.2\times10^{15} \cong \infty$$

The results for the calibration of the 30 gm UUT mass are reported as

$$x_n = 30 \text{ gm}$$

$$\bar{\delta}_{corr} = -2.5 \times 10^{-4} \text{ gm}$$

113

$$u_{cal} = 6.13 \times 10^{-2} \text{ gm , infinite degrees of freedom.}$$

8.6.2 Scenario 2: The UUT Measures the MTE Attribute Value

In this scenario, the measurement result is $\delta = x - y_n$ and ε_{cal} is expressed in equation (8-26). The example for this scenario consists of calibrating an analog micrometer with a 10 mm gage block reference. Multiple readings of the 10 mm gage block length are taken with the micrometer under laboratory environmental conditions of $24\ ^\circ\text{C} \pm 1\ ^\circ\text{C}$. The sample statistics are computed to be

Sample Mean	=	9.999 mm
Standard Deviation	=	21.7 μm
Uncertainty in Mean	=	7.7 μm
Sample Size	=	8

The measurement result is $\bar{\delta} = (9.999 - 10)\ \text{mm} = -1\ \mu\text{m}$. However, both the micrometer reading and the gage block length must be corrected to 20 °C standard reference temperature. In this example, the gage block steel has a coefficient of thermal expansion of $\alpha_{MTE} = 11.5 \times 10^{-6}/^\circ\text{C}$ and the micrometer has a coefficient of thermal expansion of $\alpha_{UUT} = 5.6 \times 10^{-6}/\ ^\circ\text{C}$. For the purposes of this analysis, the uncertainties in α_{MTE} and α_{UUT} are assumed to be negligible.

The net effect of thermal expansion on the measurement result $\bar{\delta}$ is

$$\bar{\delta}_{env} = \bar{\delta}_{UUT,env} - \bar{\delta}_{MTE,env}$$

where $\bar{\delta}_{UUT,env}$ and $\bar{\delta}_{MTE,env}$ represent thermal expansion of the micrometer and gage block dimensions, respectively. The net length expansion is computed from the temperature difference ΔT, the average measured length $\bar{x}$, the coefficient of thermal expansion for the gage block α_{MTE} and the coefficient of thermal expansion for the micrometer α_{UUT}.

$$\begin{aligned}
\bar{\delta}_{env} &= \Delta T \times \bar{x} \times (\alpha_{UUT} - \alpha_{MTE}) \\
&= 4\,^\circ\text{C} \times 9.999\,\text{mm} \times (5.6 - 11.5) \times 10^{-6}/\,^\circ\text{C} \\
&= -2.36 \times 10^{-4}\,\text{mm} = -0.236\,\mu\text{m}
\end{aligned}$$

The corrected calibration result $\bar{\delta}_{corr}$ is computed to be

$$\begin{aligned}
\bar{\delta}_{corr} &= \bar{\delta} + \bar{\delta}_{env} \\
&= -(1 + 0.236)\,\mu\text{m} \\
&= -1.24\,\mu\text{m}
\end{aligned}$$

In the micrometer calibration scenario, the following measurement process errors must be taken into account:

- Bias in the value of the 10 mm gage block length, $e_{MTE,b}$.

- Error associated with the repeat measurements taken, $e_{UUT,rep}$.
- Error associated with the analog resolution of the micrometer, $e_{UUT,res}$.
- Bias resulting from the operator's use of the micrometer to measure the gage block, $e_{UUT,op}$.
- Environmental factors error resulting from the thermal expansion correction, e_{env}.

The error in $\overline{\delta}_{corr}$ is

$$\varepsilon_{cal} = e_{MTE,b} + e_{UUT,rep} + e_{UUT,res} + e_{UUT,op} + e_{env}$$

where

$$e_{env} = e_{\Delta T} \times \overline{x}\left(\alpha_{UUT} - \alpha_{MTE}\right) = c_{\Delta T}e_{\Delta T}$$

and $c_{\Delta T}$ is the sensitivity coefficient and $e_{\Delta T}$ is the error due to the environmental temperature variation.

The uncertainty in $\overline{\delta}_{corr}$ is

$$\begin{aligned}
u_{cal} &= \sqrt{\mathrm{var}(\varepsilon_{cal})} \\
&= \sqrt{\mathrm{var}\left(e_{MTE,b}\right) + \mathrm{var}\left(e_{UUT,rep}\right) + \mathrm{var}\left(e_{UUT,res}\right) + \mathrm{var}\left(e_{UUT,op}\right) + c_{\Delta T}^2 \, \mathrm{var}\left(e_{\Delta T}\right)} \\
&= \sqrt{u_{MTE,b}^2 + u_{UUT,rep}^2 + u_{UUT,res}^2 + u_{UUT,op}^2 + c_{\Delta T}^2 u_{\Delta T}^2}.
\end{aligned}$$

The distributions, limits, confidence levels and standard uncertainties for each error source are summarized in Table 8-3.

Table 8-3. Summary of Scenario 2 Uncertainty Estimates

Error Source	Error Limits (µm)	Conf. Level (%)	Error Distribution	Degrees of Freedom	Analysis Type	Standard Uncertainty (µm)	Sensitivity Coefficient
$e_{MTE,b}$	+ 0.18, -0.13 µm	90.00	Lognormal	Infinite	B	0.09 µm	1
$e_{UUT,rep}$			Student's t	7	A	7.7 µm	1
$e_{UUT,res}$	± 5.0 µm	95.00	Normal	Infinite	B	2.6 µm	1
$e_{UUT,op}$	± 5.0 µm	95.00	Normal	Infinite	B	2.6 µm	1
$e_{\Delta T}$	± 1 °C	95.00	Normal	Infinite	B	0.51 °C	-5.9×10^{-2} µm/°C

Using the data in Table 8-3, the uncertainty in δ_{corr} is

$$\begin{aligned}
u_{cal} &= \sqrt{(0.09)^2 + (7.7)^2 + (2.6)^2 + (2.6)^2 + \left(-5.9 \times 10^{-2} \times 0.51\right)^2}\ \text{µm} \\
&= \sqrt{0.0081 + 59.29 + 6.76 + 6.76 + 0.0009}\ \text{µm} \\
&= \sqrt{72.82}\ \text{µm} = 8.53\,\text{µm}
\end{aligned}$$

The effective degrees of freedom for u_{cal} are computed using the Welch-Satterthwaite formula.

$$\nu_{eff} = \frac{u_{cal}^4}{\dfrac{u_{MTE,b}^4}{\nu_{MTE,b}} + \dfrac{u_{UUT,rep}^4}{\nu_{UUT,rep}} + \dfrac{u_{UUT,res}^4}{\nu_{UUT,res}} + \dfrac{u_{UUT,op}^4}{\nu_{UUT,op}} + \dfrac{c_{\Delta T}^4 u_{\Delta T}^4}{\nu_{\Delta T}}}$$

$$= \frac{u_{cal}^4}{\dfrac{u_{MTE,b}^4}{\infty} + \dfrac{u_{UUT,rep}^4}{7} + \dfrac{u_{UUT,res}^4}{\infty} + \dfrac{u_{UUT,op}^4}{\infty} + \dfrac{c_{\Delta T}^4 u_{\Delta T}^4}{\infty}}$$

$$= 7 \times \frac{u_{cal}^4}{u_{UUT,rep}^4}.$$

The degrees of freedom are computed to be

$$\nu_{eff} = 7 \times \left(\frac{8.53}{7.7}\right)^4 = 7 \times (1.108)^4 = 7 \times 1.507 = 10.6$$

The results for the calibration of the micrometer at 10 mm nominal length are reported as

$$y_n = 10 \text{ mm}$$
$$\bar{\delta}_{corr} = -1.24 \ \mu\text{m}$$
$$u_{cal} = 8.53 \ \mu\text{m} , \ 11 \text{ degrees of freedom}.$$

8.6.3 Scenario 3: MTE and UUT Attribute Values are Compared

In this scenario, the measurement result is $\delta = x - y$ and ε_{cal} is expressed in equation (8-39). The example for this scenario consists of calibrating an end gauge, with a nominal length of 50 mm, using an end gauge standard of the same nominal length. The calibration process consists of measuring and recording the difference between the two end gauges using a comparator apparatus.

In this case, the difference in the lengths of the two end gauges are measured. The sample statistics are computed to be

Sample Mean	=	215 nm
Standard Deviation	=	9.7 nm
Uncertainty in Mean	=	4.33 nm
Sample Size	=	5

and the measurement result is $\bar{\delta} = 215 \text{ nm}$. The temperature for both end gauges during calibration is 19.9 °C ± 0.5 °C. Consequently, the calibration result must be corrected to the standard reference temperature of 20 °C. The corrected calibration result $\bar{\delta}_{corr}$ is computed from

116

$$\bar{\delta}_{corr} = \bar{\delta} + \bar{\delta}_{env}$$
$$= \bar{\delta} + \bar{\delta}_{UUT,env} - \bar{\delta}_{MTE,env}$$

where

$$\bar{\delta}_{UUT,env} = \Delta T \times \bar{x} \times \alpha_{UUT} = \text{thermal expansion of the UUT end gage}$$

$$\bar{\delta}_{MTE,env} = \Delta T \times \bar{y} \times \alpha_{MTE} = \text{thermal expansion of the MTE end gage}$$

$$\bar{x} = \text{the average UUT end gage length during calibration}$$
$$\bar{y} = \text{the average MTE end gage length during calibration}$$
$$\alpha_{UUT} = \text{coefficient of thermal expansion for the UUT end gauge}$$
$$\alpha_{MTE} = \text{coefficient of thermal expansion for the MTE end gauge}$$
$$\Delta T = \text{difference in the temperature of the end gauge from the 20 °C}$$

For the purposes of this example, we will assume that $\alpha_{UUT} = \alpha_{MTE} = \alpha = 11.5 \times 10^{-6}/°C$. Therefore, $\bar{\delta}_{corr}$ can be expressed as

$$\bar{\delta}_{corr} = \bar{\delta} + \Delta T \alpha \left(\bar{x} - \bar{y} \right) = \bar{\delta} + \Delta T \alpha \bar{\delta}$$
$$= \bar{\delta} \left(1 + \Delta T \alpha \right)$$

and is computed to be

$$\bar{\delta}_{corr} = 215 \, \text{nm} \left(1 - 0.1°C \times 11.5 \times 10^{-6} / °C \right)$$
$$= 215 \, \text{nm} \left(1 - 1.15 \times 10^{-6} \right)$$
$$\cong 215 \, \text{nm}.$$

In the end gauge calibration scenario, the following measurement process errors must be taken into account:

- Bias in the value of the 50 mm end gage standard length, $e_{MTE,b}$.
- Bias of the comparator, $e_{c,b}$
- Error associated with the repeat measurements taken, e_{rep}.
- Digital resolution error for the comparator, e_{res}.
- Operator bias, e_{op}
- Environmental factors error resulting from the thermal expansion correction, e_{env}.

As shown in equations (8-37) through (8-39), the comparator bias cancels out and the error in $\bar{\delta}_{corr}$ is

$$\varepsilon_{cal} = e_{rep} + e_{res} + e_{op} + e_{env} - e_{MTE,b}$$

where

$$e_{rep} = e_{UUT,rep} - e_{MTE,rep} = e_{\bar{\delta},rep}$$

$$e_{res} = e_{UUT,res} - e_{MTE,res}$$

$$e_{env} = e_{UUT,env} - e_{MTE,env} = e_{\bar{\delta},env}$$

and

$$e_{\bar{\delta},env} = c_1 e_{\Delta T} + c_2 e_\alpha.$$

The uncertainty in $\bar{\delta}_{corr}$ is

$$u_{cal} = \sqrt{\mathrm{var}(\varepsilon_{cal})}$$

$$= \sqrt{\mathrm{var}\left(e_{\bar{\delta},rep}\right) + \mathrm{var}\left(e_{UUT,res} - e_{MTE,res}\right) + \mathrm{var}\left(c_1 e_{\Delta T} + c_2 e_\alpha\right) + \mathrm{var}\left(e_{op}\right) + \mathrm{var}\left(-e_{MTE,b}\right)}$$

$$= \sqrt{u_{\bar{\delta},rep}^2 + u_{UUT,res}^2 + u_{MTE,res}^2 - 2\rho_{res} u_{UUT,res} u_{MTE,res} + c_1^2 u_{\Delta T}^2 + c_2^2 u_\alpha^2 + 2c_1 c_2 \rho_{env} u_{\Delta T} u_\alpha + u_{op}^2 + u_{MTE,b}^2}.$$

The UUT and MTE resolution uncertainties are equal to the comparator resolution uncertainty, $u_{UUT,res} = u_{MTE,res} = u_{c,res}$. In addition, the UUT and MTE resolution errors are uncorrelated, so that $\rho_{res} = 0$. The UUT and MTE end gage length expansion corrections will err in the same direction and by a constant proportional amount, so that $\rho_{env} = 1$. Therefore, the uncertainty u_{cal} can be expressed as

$$u_{cal} = \sqrt{u_{\bar{\delta},rep}^2 + 2u_{c,res}^2 + c_1^2 u_{\Delta T}^2 + c_2^2 u_\alpha^2 + 2c_1 c_2 u_{\Delta T} u_\alpha + u_{op}^2 + u_{MTE,b}^2}$$

$$= \sqrt{u_{\bar{\delta},rep}^2 + 2u_{c,res}^2 + \left(c_1 u_{\Delta T} + c_2 u_\alpha\right)^2 + u_{MTE,b}^2}$$

The sensitivity coefficients c_1 and c_2 are

$$c_1 = \frac{\partial \bar{\delta}_{env}}{\partial \Delta T} = \alpha \bar{\delta} = 11.5 \times 10^{-6} / {}^\circ C \times 215 \, nm \quad \text{and} \quad c_2 = \frac{\partial \bar{\delta}_{env}}{\partial \alpha} = \Delta T \bar{\delta} = -0.1 {}^\circ C \times 215 \, nm$$

$$= 2.47 \times 10^{-3} \, nm / {}^\circ C \qquad\qquad = -21.5 {}^\circ C \bullet nm$$

Using the data in Table 8-4, the uncertainty in $\bar{\delta}_{corr}$ is

The distributions, limits, confidence levels and standard uncertainties for each error source are summarized in Table 8-4.

Table 8-4. Summary of Scenario 3 Uncertainty Estimates

Error Source	Error Limits	Conf. Level (%)	Error Distribution	Degrees of Freedom	Analysis Type	Standard Uncertainty (nm)	Sensitivity Coefficient
$e_{\bar{\delta},rep}$			Student's t	4	A	4.33 nm	1
$e_{c,res}$	± 1 nm	100.0	Uniform	Infinite	B	0.577 nm	1
$e_{\Delta T}$	$\pm 0.5 \,{}^\circ C$	95.00	Normal	Infinite	B	$0.255 \,{}^\circ C$	2.47×10^{-3} nm/${}^\circ C$
e_α	$\pm 0.5 \times 10^{-6} /{}^\circ C$	95.00	Normal	Infinite	B	$0.255 \times 10^{-6} /{}^\circ C$	$-21.5 \,{}^\circ C \bullet nm$
e_{op}	± 5 nm	95.00	Normal	Infinite	B	2.55 nm	1
$e_{MTE,b}$			Normal	18	A,B	25 nm	1

118

$$u_{cal} = \sqrt{(4.33)^2 + 2 \times (0.577)^2 + (2.47 \times 10^{-3} \times 0.255 - 21.5 \times 0.255 \times 10^{-6})^2 + (2.55)^2 + (25)^2} \text{ nm}$$

$$= \sqrt{18.75 + 0.67 + 3.90 \times 10^{-7} + 6.51 + 625} \text{ nm}$$

$$= \sqrt{650.9} \text{ nm} = 25.5 \text{ nm}.$$

The effective degrees of freedom for u_{cal} are computed using the Welch-Satterthwaite formula.

$$\nu_{eff} = \frac{u_{cal}^4}{\dfrac{u_{\bar{\delta},rep}^4}{\nu_{\bar{\delta},rep}} + 2\dfrac{u_{c,res}^4}{\nu_{c,res}} + \dfrac{c_1^4 u_{\Delta T}^4}{\nu_{\Delta T}} + \dfrac{c_2^4 u_\alpha^4}{\nu_\alpha} + \dfrac{u_{MTE,b}^4}{\nu_{MTE,b}}}$$

$$= \frac{u_{cal}^4}{\dfrac{u_{\bar{\delta},rep}^4}{5} + 2\dfrac{u_{c,res}^4}{\infty} + \dfrac{c_1^4 u_{\Delta T}^4}{\infty} + \dfrac{c_2^4 u_\alpha^4}{\infty} + \dfrac{u_{MTE,b}^4}{18}} = \frac{u_{cal}^4}{\dfrac{u_{\bar{\delta},rep}^4}{5} + \dfrac{u_{MTE,b}^4}{18}}.$$

The degrees of freedom are computed to be

$$\nu_{eff} = \frac{(25.5)^4}{\dfrac{(4.33)^4}{5} + \dfrac{(25)^4}{18}} = \frac{423,680.8}{70.3 + 21,701.4} = \frac{423,680.8}{21,771.7} = 19.5$$

The results for the calibration of the UUT end gauge with 50 mm nominal length are reported as

$$x_n = 50 \text{ mm}$$
$$\bar{\delta}_{corr} = 215 \text{ nm}$$
$$u_{cal} = 25.4 \text{ nm} \text{ , 20 degrees of freedom.}$$

8.6.4   Scenario 4: The MTE and UUT Measure a Common Artifact

For this scenario, both the MTE and UUT measure the value or output of a common artifact. The measurement result is $\delta = x - y$ and ε_{cal} is expressed in equation (8-48).

The example for this scenario consists of calibrating a digital thermometer at 100 °C using an oven and an analog temperature reference. The oven temperature is adjusted using its internal temperature probe and the readings from the thermometer and temperature reference are recorded. The resulting sample statistics are computed to be

Sample Mean, UUT	= 100.50 °C	Sample Mean, MTE	= 100.000 °C
Standard Deviation, UUT	= 0.03 °C	Standard Deviation, MTE	= 0.006 °C
Uncertainty in Mean, UUT	= 0.01 °C	Uncertainty in Mean, MTE	= 0.002 °C
Sample Size	= 9	Sample Size	= 9

and the measurement result is $\bar{\delta} = 100.50 - 100.000 = 0.50 \,°C$. In the thermometer calibration scenario, the following measurement process errors must be taken into account:

- Bias of the temperature reference, $e_{MTE,b}$.
- Error due to repeat measurements taken with the temperature reference, $e_{MTE,rep}$.
- Error due to repeat measurements taken with the thermometer, $e_{UUT,rep}$.
- Analog resolution error for the temperature reference, $e_{MTE,res}$.
- Digital resolution error for the thermometer, $e_{UUT,res}$.
- Error due to the non-uniformity of the oven temperature, e_{env}.

The short-term effect of oven stability is accounted for in the sample of MTE measurements. If repeat measurements were not collected, then the error due to oven stability would be included as part of the environmental factors error, e_{env}.

The error in $\bar{\delta}$ is

$$\varepsilon_{cal} = \left(e_{UUT,rep} - e_{MTE,rep}\right) + \left(e_{UUT,res} - e_{MTE,res}\right) - e_{MTE,b} + e_{env}.$$

The uncertainty in $\bar{\delta}$ is

$$\begin{aligned}
u_{cal} &= \sqrt{\mathrm{var}(\varepsilon_{cal})} \\
&= \sqrt{\mathrm{var}(-e_{MTE,b}) + \mathrm{var}(e_{UUT,rep} - e_{MTE,rep}) + \mathrm{var}(e_{UUT,res} - e_{MTE,res}) + \mathrm{var}\left(e_{env}\right)} \\
&= \sqrt{u_{MTE,b}^2 + u_{UUT,rep}^2 + u_{MTE,rep}^2 + u_{UUT,res}^2 + u_{MTE,res}^2 + u_{env}^2}.
\end{aligned}$$

The distributions, limits, confidence levels and standard uncertainties for each error source are summarized in Table 8-5.

Table 8-5. Summary of Scenario 4 Uncertainty Estimates

Error Source	Error Limits (°C)	Confidence Level (%)	Error Distribution	Degrees of Freedom	Analysis Type	Standard Uncertainty (°C)
$e_{MTE,b}$			Normal	29	A,B	0.02
$e_{UUT,rep}$			Student's t	8	A	0.22
$e_{MTE,rep}$			Student's t	8	A	0.075
$e_{UUT,res}$	± 0.005	100.00	Uniform	infinite	B	0.0029
$e_{MTE,res}$	± 0.0025	95.00	Normal	infinite	B	0.0013
e_{env}	± 2	95.00	Normal	Infinite	B	1.02

Using the data in Table 8-5, the uncertainty in $\bar{\delta}$ is

$$\begin{aligned}
u_{cal} &= \sqrt{\left(0.02\right)^2 + \left(0.22\right)^2 + \left(0.075\right)^2 + \left(0.0029\right)^2 + \left(0.0013\right)^2 + \left(1.02\right)^2}\ °\mathrm{C} \\
&= \sqrt{0.0004 + 0.0484 + 0.0056 + 0.000008 + 0.000002 + 1.0404}\ °\mathrm{C} \\
&= \sqrt{1.095}\ °\mathrm{C} = 1.05\ °\mathrm{C}
\end{aligned}$$

The Welch-Satterthwaite formula for the effective degrees of freedom for u_{cal} is

$$v_{eff} = \frac{u_{cal}^4}{\dfrac{u_{MTE,b}^4}{v_{MTE,b}} + \dfrac{u_{UUT,rep}^4}{v_{UUT,rep}} + \dfrac{u_{MTE,rep}^4}{v_{MTE,rep}} + \dfrac{u_{UUT,res}^4}{v_{UUT,res}} + \dfrac{u_{MTE,res}^4}{v_{MTE,res}} + \dfrac{u_{env}^4}{v_{env}}}$$

$$= \frac{u_{cal}^4}{\dfrac{u_{MTE,b}^4}{29} + \dfrac{u_{UUT,rep}^4}{8} + \dfrac{u_{MTE,rep}^4}{8} + \dfrac{u_{UUT,res}^4}{\infty} + \dfrac{u_{MTE,res}^4}{\infty} + \dfrac{u_{env}^4}{\infty}}$$

$$= \frac{u_{cal}^4}{\dfrac{u_{MTE,b}^4}{29} + \dfrac{u_{UUT,rep}^4}{8} + \dfrac{u_{MTE,rep}^4}{8}}.$$

The degrees of freedom are computed to be

$$v_{eff} = \frac{(1.05)^4}{\dfrac{(0.02)^4}{29} + \dfrac{(0.22)^4}{8} + \dfrac{(0.075)^4}{8}}$$

$$= \frac{1.22}{5.52 \times 10^{-9} + 2.93 \times 10^{-4} + 3.96 \times 10^{-6}}$$

$$= \frac{1.22}{2.97 \times 10^{-4}} = 4095.6$$

The results for the calibration of the UUT digital thermometer at 100 °C nominal temperature are reported as

$$\begin{aligned}
\bar{x} &= 100.50\ °C \\
\bar{y} &= 100.000\ °C \\
\bar{\delta} &= 0.50\ °C \\
u_{cal} &= 1.05\ °C, \text{ infinite degrees of freedom.}
\end{aligned}$$

CHAPTER 9: UNCERTAINTY GROWTH

Over time, the error or bias in an MTE attribute or parameter may increase, remain constant or decrease. The uncertainty in this error, however, *always* increases with time since measurement or calibration. This is the **fundamental postulate** of uncertainty growth. This chapter discusses the methodology to project the growth in the MTE attribute or parameter bias uncertainty.[63]

Figure 9-1 illustrates uncertainty growth over time for a typical attribute or parameter bias, ε_b. The sequence shows the probability distribution at three different times, with the uncertainty growth reflected in the spreads in the curves. The out-of-tolerance probabilities at the different times are represented by the shaded areas under the curves.

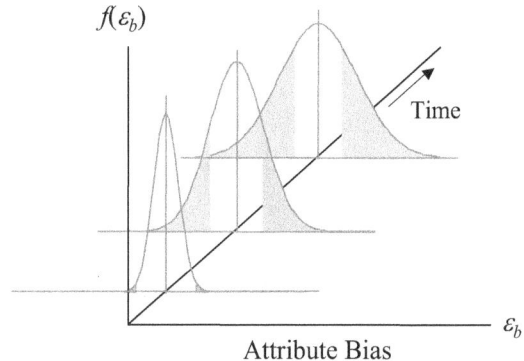

Figure 9-1. Measurement Uncertainty Growth

Uncertainty growth over time corresponds to an increase in out-of-tolerance probability over time, or equivalently, to a decrease in in-tolerance probability or measurement reliability $R(t)$ over time. Plotting $R(t)$ versus time, as shown in Figure 9-2, suggests that measurement reliability can be modeled by a time-varying function. Once this function is determined, then MTE parameter bias uncertainty can be computed as a function of time.

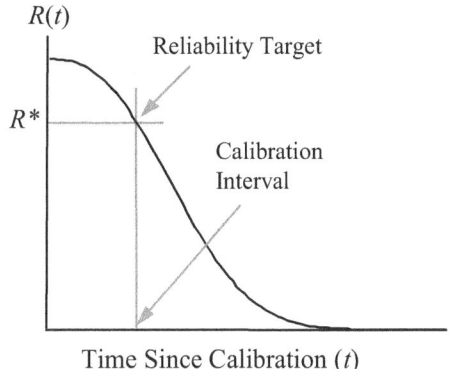

Figure 9-2. Measurement Reliability versus Time

9.1 Basic Methodology

The uncertainty, $u(t)$, in the parameter bias, $\varepsilon_b(t)$, at time t elapsed since measurement is

---

[63] Dr. H. Castrup, "Uncertainty Growth Estimation in UncertaintyAnalyzer," Integrated Sciences Group Technical Report, March 27, 2002. Revised June 17, 2008.

computed using the value of the initial measurement uncertainty and the reliability model for the parameter population. The basic concept is an extension of the ergodic theorem[64] that states that the distribution of an infinite population of values at equilibrium is identical to the distribution of values attained by a single member sampled an infinite number of times.

The reliability model predicts the in-tolerance probability for the parameter bias population as a function of time elapsed since measurement. It can be thought of as a function that quantifies the *stability* of the population. In this view, we begin with a population measurement reliability immediately following measurement at time $t = 0$ and extrapolate to the measurement reliability at time $t > 0$. The measurement reliability of the parameter bias at time t is related to the bias uncertainty according to

$$R(t) = \int_{-L_1}^{L_2} f[\varepsilon_b(t)]d\varepsilon_b \qquad (9\text{-}1)$$

where $f[\varepsilon_b(t)]$ is the probability density function (pdf) for the parameter bias and $-L_1$ and L_2 are the tolerance limits. For example, if we assume that $\varepsilon_b(t)$ is normally distributed, then

$$f[\varepsilon_b(t)] = \frac{1}{\sqrt{2\pi}u(t)} e^{-[\varepsilon_b(t)-\mu(t)]^2 / 2u^2(t)} \qquad (9\text{-}2)$$

where $\mu(t)$ represents the expected or true parameter bias at time t. The relationship between L_1, L_2, $\varepsilon_b(t)$ and $\mu(t)$ is shown in Figure 9-3, along with the distribution of the population of biases for the measurement parameter of interest.

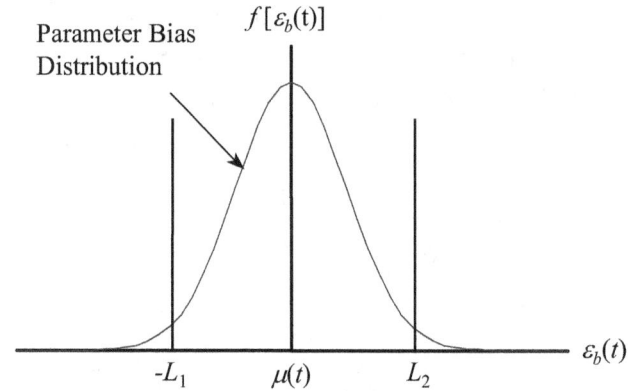

Figure 9-3. Parameter Bias Distribution

At time t, $\mu(t)$ is defined by the relation

$$\mu(t) = \mu_0 + \alpha(t) \qquad (9\text{-}3)$$

[64] See for example, Gray, R. M.: *Probability, Random Processes, and Ergodic Properties*, Springer-Verlag 1987. Revised 2001 and 2006-2007 by Robert M. Gray.

where μ_0 is the true parameter bias at $t = 0$ and $\alpha(t)$ is a drift function.[65]

At the time of measurement ($t = 0$), a value for μ_0 is estimated (measured) and the uncertainty in this estimate is designated u_0. The following section discusses a method for calculating $u(t)$, given u_0.

9.2 Projected Uncertainty

Given the reliability model for the individual parameter bias and its initial uncertainty, the uncertainty $u(t)$ in equation (9-1) could be obtained directly by iteration or other means. However, available information typically relates only to the characteristics of the reliability model for the population to which the parameter belongs. Therefore, the reliability model for the population is applied.

Using a population reliability model to estimate uncertainty growth for an individual parameter employs the following set of assumptions:

1. The result of a measurement is an estimate of a parameter's value or bias. This result is accompanied by an estimate of the uncertainty in the parameter's bias.

2. The uncertainty of the measured parameter's bias or value at time $t = 0$ (immediately following measurement) is the estimated uncertainty of the measurement process.

3. The measured parameter bias or value is normally distributed around the measurement result.

4. The stability of the parameter is equated to the stability of its population. This stability is represented by the population's reliability model.

5. Therefore, the uncertainty in the parameter's bias or value grows from its value at $t = 0$ in accordance with the reliability model of the parameter's population.

The expressions used to compute uncertainty growth vary depending on whether the parameter tolerances are two-sided, single-sided upper or single-sided lower.

9.2.1 Two-Sided Tolerances

From equations (9-1) and (9-2), the reliability function for parameters with two-sided tolerance limits is given by

$$R(t) = \frac{1}{\sqrt{2\pi}u(t)} \int_{-L_1}^{L_2} e^{-[\varepsilon_b - \mu(t)]^2 / 2u^2(t)} d\varepsilon_b . \qquad (9\text{-}4)$$

We define a variable ζ as

$$\zeta \equiv \frac{\varepsilon_b(t) - \mu(t)}{u(t)}$$

---

[65] The drift function can be a linear or other mathematical relation. The only restriction is that at $t = 0$, $\alpha(t) = 0$.

so that $d\varepsilon_b = u(t)d\zeta$. Using substitution of variables, equation (9-4) becomes

$$R(t) = \frac{1}{\sqrt{2\pi}} \int_{-[L_1+\mu(t)]/u(t)}^{[L_2-\mu(t)]/u(t)} e^{-\zeta^2/2}d\zeta$$

$$= \Phi\left(\frac{L_1 + \mu(t)}{u(t)}\right) + \Phi\left(\frac{L_2 - \mu(t)}{u(t)}\right) - 1 \qquad (9\text{-}5)$$

where the function $\Phi(\cdot)$ is defined by

$$\Phi(z) = \frac{1}{\sqrt{2\pi}} \int_{-\infty}^{z} e^{-\zeta^2/2}d\zeta \,.$$

At $t = 0$, the reliability function is

$$R(0) = \Phi\left(\frac{L_1 + \mu(0)}{u(0)}\right) + \Phi\left(\frac{L_2 - \mu(0)}{u(0)}\right) - 1 \qquad (9\text{-}6)$$

where $\mu(0)$ is set equal to the mean or average value obtained from a sample of measurements or to the value obtained as a Bayesian estimate.[66]

If $\mu(0)$ is set equal to a sample mean value, u_0 is set equal to the combined uncertainty estimate for the mean value. If $\mu(0)$ is set equal to a Bayesian estimate, $u(0)$ is set equal to the uncertainty of the Bayesian estimate.

Equations (9-5) and (9-6) are used to estimate uncertainty growth. Since this growth consists of an increase in the initial uncertainty estimate, based on knowledge of the stability of the parameter population, it should not be influenced by the quantity $\mu(t)$. Accordingly, two reliability functions, R_0 and R_t, are defined by

$$R_0 = \Phi\left(\frac{L_1}{u_0{}'}\right) + \Phi\left(\frac{L_2}{u_0{}'}\right) - 1 \qquad (9\text{-}7)$$

and

$$R_t = \Phi\left(\frac{L_1}{u_t{}'}\right) + \Phi\left(\frac{L_2}{u_t{}'}\right) - 1 \,. \qquad (9\text{-}8)$$

where R_0 and R_t are computed from the population reliability models at times 0 and t, respectively.

Next, the variables $u_0{}'$ and $u_t{}'$ are solved for iteratively using the bisection method.[67] The

---

[66] Bayesian analysis is discussed in Appendix E.

[67] See Chapter 9 of Press, et al., *Numerical Recipes in Fortran*, 2nd Ed., Cambridge University Press, 1992.

solutions are used to obtain $u(t)$ from the following relation

$$u(t) = u(0) \frac{u_t{}'}{u_0{}'}.$$

(9-9)

The in-tolerance probability at time t is then solved for using equation (9-5). A "best" estimate for μ is obtained using equation (9-3). If the function $\alpha(t)$ is not known, the last known value of μ, namely $\mu(0)$, is used. If this is the case, substituting $\mu(0)$ for μ in equation (9-5) gives an estimate of the in-tolerance probability at time $t > 0$.

$$R(t) \cong \Phi\left[\frac{L_1 + \mu(0)}{u(t)}\right] + \Phi\left[\frac{L_2 - \mu(0)}{u(t)}\right] - 1$$

(9-10)

9.2.2 Two-Sided Symmetric Tolerances

If the tolerance limits are symmetric, then $L_1 = L_2 = L$ and equations (9-7) through (9-10) are applied. In cases where $\mu_0 = 0$, then equations (9-7) and (9-8) become

$$R_0 = 2\Phi\left(\frac{L}{u_0{}'}\right) - 1$$

(9-11)

and

$$R_t = 2\Phi\left(\frac{L}{u_t{}'}\right) - 1.$$

(9-12)

The variables $u_0{}'$ and $u_t{}'$ are

$$u_0{}' = \frac{L}{\Phi^{-1}\left(\dfrac{1 + R_0}{2}\right)}$$

(9-13)

and

$$u_t{}' = \frac{L}{\Phi^{-1}\left(\dfrac{1 + R_t}{2}\right)}.$$

(9-14)

Applying equation (9-9), $u(t)$ is computed from

$$u(t) = u(0) \frac{\Phi^{-1}\left(\dfrac{1 + R_0}{2}\right)}{\Phi^{-1}\left(\dfrac{1 + R_t}{2}\right)}.$$

(9-15)

The in-tolerance probability at time $t > 0$ is

$$R(t) \cong 2\Phi\left(\frac{L}{u(t)}\right) - 1.$$

(9-16)

126

9.2.3 Single-Sided Tolerances

In cases where tolerances are single-sided, either L_1 or L_2 is infinite. For **single-sided lower limit** cases, $L_1 = L$, $L_2 = \infty$ and equation (9-5) becomes

$$
\begin{aligned}
R(t) &= \Phi\left(\frac{L + \mu(t)}{u(t)}\right) + \Phi\left(\frac{\infty - \mu(t)}{u(t)}\right) - 1 \\
&= \Phi\left(\frac{L + \mu(t)}{u(t)}\right) + \Phi(\infty) - 1 \\
&= \Phi\left(\frac{L + \mu(t)}{u(t)}\right) + 1 - 1 \\
&= \Phi\left(\frac{L + \mu(t)}{u(t)}\right).
\end{aligned}
\tag{9-17}
$$

Equation (9-6) similarly becomes

$$
R(0) = \Phi\left(\frac{L + \mu(0)}{u(0)}\right).
\tag{9-18}
$$

For **single-sided upper limit** cases, $L_1 = \infty$, $L_2 = L$ and equations (9-5) and (9-6) become

$$
\begin{aligned}
R(t) &= \Phi\left(\frac{\infty + \mu(t)}{u(t)}\right) + \Phi\left(\frac{L - \mu(t)}{u(t)}\right) - 1 \\
&= \Phi\left(\frac{L - \mu(t)}{u(t)}\right)
\end{aligned}
\tag{9-19}
$$

and

$$
R(0) = \Phi\left(\frac{L - \mu(0)}{u(0)}\right).
\tag{9-20}
$$

For both single-sided upper limit and single-sided lower limit cases, equations (9-7) and (9-8) become

$$
R_0 = \Phi\left(\frac{L}{u_0'}\right)
\tag{9-21}
$$

and

$$
R_t = \Phi\left(\frac{L}{u_t'}\right).
\tag{9-22}
$$

The variables u_0' and u_t' are

127

$$u_0' = \frac{L}{\Phi^{-1}(R_0)} \tag{9-23}$$

and

$$u_t' = \frac{L}{\Phi^{-1}(R_t)}. \tag{9-24}$$

Applying equation (9-9), $u(t)$ is computed from

$$u(t) = u(0)\frac{\Phi^{-1}(R_0)}{\Phi^{-1}(R_t)}. \tag{9-25}$$

Applying equation (9-10), the in-tolerance probability at time t is

$$R(t) \cong \Phi\left(\frac{L + \mu(0)}{u(t)}\right). \tag{9-26}$$

9.3 Reliability Models

In the uncertainty growth projection process, information about the calibration history of the parameter population is used to develop a reliability model. This reliability model provides a means for determining how the parameter bias uncertainty grows with time since calibration.

Each reliability model is defined by a mathematical equation with coefficients. A calibration interval analysis program can be used to determine the reliability model that "best fits" a parameter's calibration history data and to compute the corresponding model coefficients. If a reliability modeling application is not accessible, then an applicable reliability model must be chosen based on knowledge about the stability of the subject parameter over time.

Commonly used reliability models are described in the following subsections along with information needed to implement them. Guidance on the selection and application of these reliability models can be found in NCSL RP-1 *Establishment and Adjustment of Calibration Intervals*.[68]

9.3.1 Exponential Model

The exponential reliability model is defined by the mathematical equation

$$R(t) = ae^{-bt} \tag{9-27}$$

where a and b are the model coefficients. An example plot for the exponential model is shown in Figure 9-4.

The exponential model is useful for parameters whose failure probability is not a function of time since last measurement. That is, the probability of going out-of-tolerance in the interval

---

[68] NCSL, *Establishment and Adjustment of Calibration Intervals*, Recommended Practice RP-1, National Conference of Standards Laboratories, January 1996.

T, beginning at some time *t*, is the same as the probability of going out-of-tolerance in the same time interval *T*, beginning at some other time *t'*.

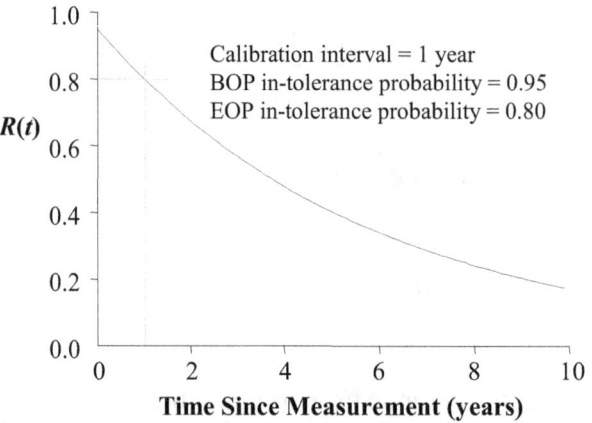

Figure 9-4. In-tolerance Probability versus Time – Exponential Model

To implement the exponential model, either of the following must be known:

1. The value of the model coefficients, *a* and *b*.
2. The beginning of period (BOP) in-tolerance probability and the end of period (EOP) in-tolerance probability.

9.3.2 Mixed Exponential Model

The mixed exponential reliability model is defined by the mathematical equation

$$R(t) = \frac{1}{\left(1 + \dfrac{at}{b}\right)^b} \tag{9-28}$$

where *a* and *b* are the model coefficients. An example plot for the exponential model is shown in Figure 9-5.

The mixed exponential model is useful for parameters whose out-of-tolerance behavior depends on a number of constituent parameters, each of which can be represented with the exponential model and where the coefficient *b* is gamma distributed.

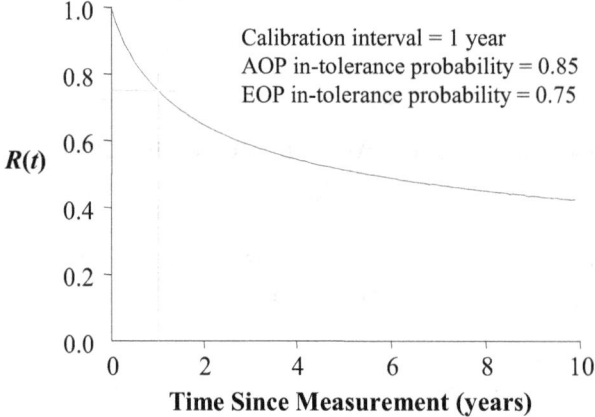

Figure 9-5. In-tolerance Probability versus Time – Mixed Exponential Model

To implement the mixed exponential model, either of the following must be known:

1. The value of the model coefficients, *a* and *b*.

2. The average over period (AOP) and EOP in-tolerance probabilities.

9.3.3   Weibull Model

The Weibull reliability model is defined by the mathematical equation

$$R(t) = ae^{-(bt)^c} \tag{9-29}$$

where *a*, *b* and *c* are the model coefficients. An example plot for the Weibull model is shown in Figure 9-6.

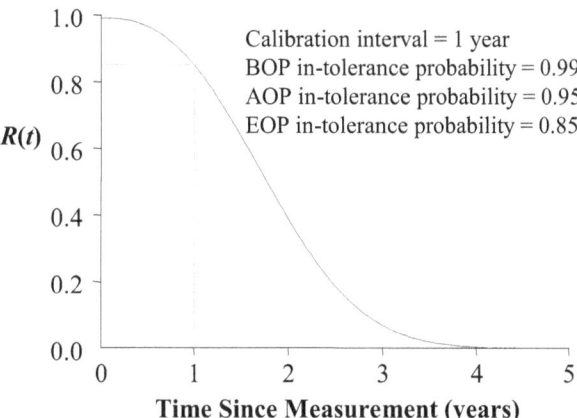

Figure 9-6. **In-tolerance Probability versus Time – Weibull Model**

The Weibull model is useful for parameters that go out-of-tolerance as a result of gradual wear or decay.

To implement the Weibull model, either of the following must be known:

1. The value of the model coefficients, *a*, *b* and *c*.

2. The BOP, AOP and EOP in-tolerance probabilities.

9.3.4   Gamma Model

The gamma reliability model is defined by the mathematical equation

$$R(t) = ae^{-bt}\left[1 + bt + \frac{(bt)^2}{2} + \frac{(bt)^3}{6}\right] \tag{9-30}$$

where *a* and *b* are the model coefficients. An example plot for the gamma model is shown in Figure 9-7.

The gamma model is useful for parameters that go out-of-tolerance in response to some number of events, such as being activated and deactivated.

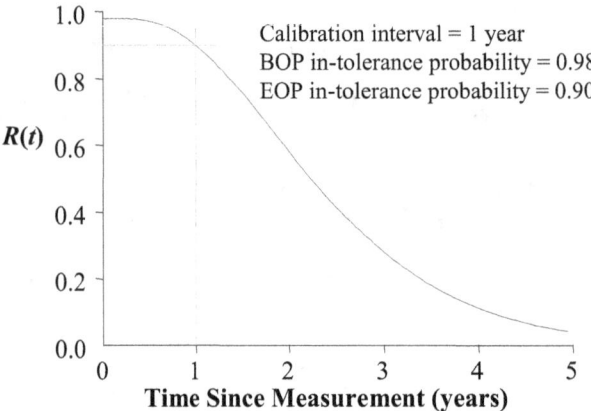

Figure 9-7. In-tolerance Probability versus Time – Gamma Model.

To implement the gamma model, either of the following must be known:

1. The value of the model coefficients, *a* and *b*.

2. The BOP and EOP in-tolerance probabilities.

9.3.5   Mortality Drift Model

The mortality drift reliability model is defined by the mathematical equation

$$R(t) = ae^{-\left(bt + ct^2\right)}$$ (9-31)

where *a*, *b* and *c* are the model coefficients. An example plot for the mortality drift model is shown in Figure 9-8.

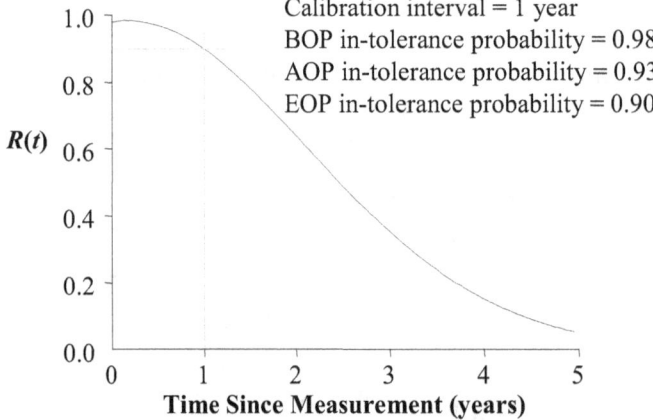

Figure 9-8. In-tolerance Probability versus Time – Mortality Drift Model

The mortality drift model is useful for parameters that are characterized by a slowly-varying out-of-tolerance rate.

To implement the mortality drift model you need to know either of the following:

1. The value of the model coefficients, a, b and c.
2. The BOP, AOP, and EOP in-tolerances.

9.3.6 Warranty Model

The warranty reliability model is defined by the mathematical equation

$$R(t) = \frac{1}{1 + e^{a(t-b)}} \tag{9-32}$$

where a and b are the model coefficients. An example plot for the warranty model is shown in Figure 9-9.

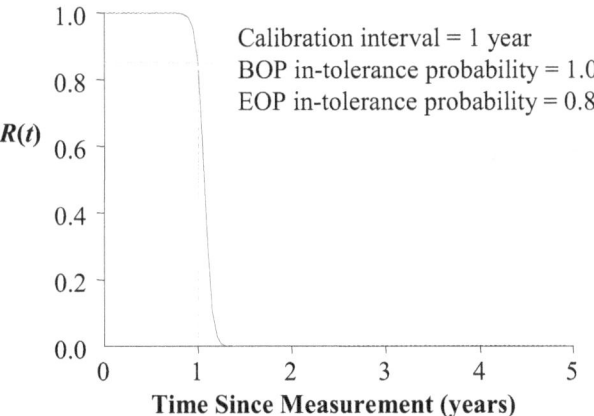

Calibration interval = 1 year
BOP in-tolerance probability = 1.00
EOP in-tolerance probability = 0.85

Figure 9-9. In-tolerance Probability versus Time – Warranty Model

The warranty model is useful for parameters that tend to stay in-tolerance until reaching a well-defined cut-off time, at which point, they go out-of-tolerance.

To implement the warranty model, one of the following must be known:

1. The value of the model coefficients, a and b.
2. The BOP and EOP in-tolerance probabilities.

9.3.7 Random Walk Model

The random walk reliability model is defined by the mathematical equation

$$R(t) = erf\left(\frac{1}{\sqrt{a + bt}}\right) \tag{9-33}$$

where erf is the error function[69] and a and b are the model coefficients. An example plot for the

---

[69] See Chapter 7 of Abramowitz, M. and Stegun, I. A., *Handbook of Mathematical Functions with Formulas, Graphs, and Mathematical Tables*, National Bureau of Standards, 1970. This handbook is eventually to be replaced by the NIST Digital Library of Mathematical Functions (http://dlmf.nist.gov/) which is currently under development.

random walk model is shown in Figure 9-10.

The random walk model is useful for parameters whose values fluctuate in a purely random way with respect to magnitude and direction (positive or negative).

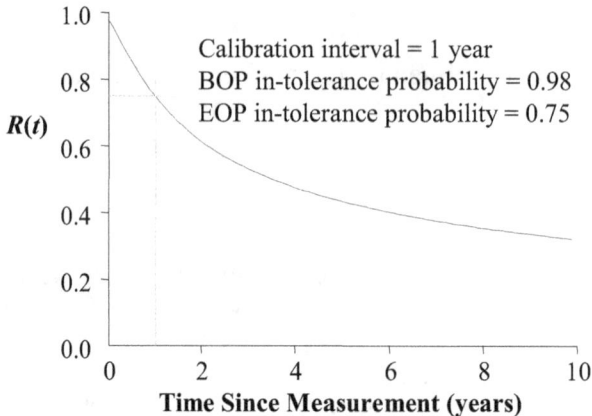

Figure 9-10. In-tolerance Probability versus Time – Random Walk Model

To implement the random walk model, one of the following must be known:

1. The value of the model coefficients, a and b.
2. The BOP and EOP in-tolerance probabilities.

9.3.8   Restricted Random Walk Model

The restricted random walk reliability model is defined by the mathematical equation

$$R(t) = erf\left(\frac{1}{\sqrt{a + b\left(1 - e^{-ct}\right)}}\right) \tag{9-34}$$

where a, b and c are the model coefficients. An example plot for the restricted random walk model is shown in Figure 9-11.

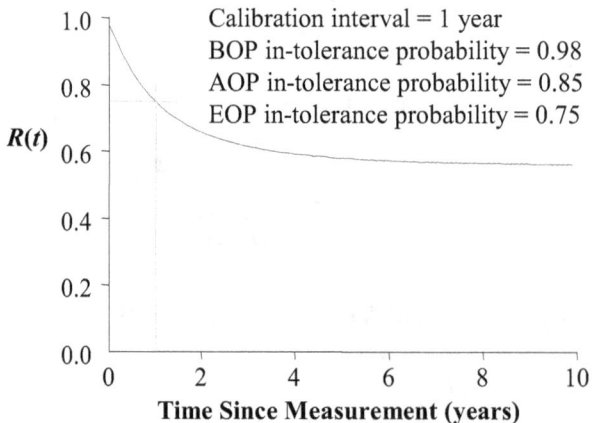

Figure 9-11. In-tolerance Probability versus Time – Restricted Random Walk Model

133

The restricted random walk model is similar to the random walk model, except that parameter fluctuations are confined within a restricted region around a mean or nominal value.

To implement the restricted random walk model, one of the following must be known:

1. The value of the model coefficients, a, b and c.
2. The BOP, AOP, and EOP in-tolerances.

9.4   Analysis Example

For illustrative purposes, let us consider an MTE attribute or parameter that comes from a population whose in-tolerance probability over time can be characterized by the gamma model described in section 9.3.4. For this example, the MTE parameter is calibrated annually and the BOP and EOP reliabilities are 0.98 and 0.90, respectively.

From equation (9-30), expressing time t in units of years,

$$R_0 = R(0) = ae^0[1] = a = 0.98 \quad \text{and} \quad R_t = R(1) = 0.98e^{-b}\left[1 + b + \frac{b^2}{2} + \frac{b^3}{6}\right] = 0.90 .$$

From iteration, the coefficient b is found to be equal to 1.62.

Now, let us assume that the MTE parameter is a 10 VDC output from a multifunction calibrator. We will also assume that the parameter tolerance limits are ± 5 μV. During calibration, the parameter bias uncertainty $u(0)$ is estimated to be 1.5 μV.

Applying the gamma model, we can estimate the uncertainty growth 6 months (0.5 years) after calibration. In this case, $t = 0.5$ years, $bt = 1.62 \times 0.5 = 0.81$ and the in-tolerance probability is computed to be

$$R_t = R(0.5) = 0.98e^{-0.81}\left[1 + 0.81 + \frac{(0.81)^2}{2} + \frac{(0.81)^3}{6}\right]$$

$$= 0.98 \times 0.44486 \times \left[1 + 0.81 + \frac{0.6561}{2} + \frac{0.5314}{6}\right]$$

$$= 0.98 \times 0.44486 \times [1 + 0.81 + 0.3281 + 0.0886]$$

$$= 0.98 \times 0.44486 \times 2.2266 = 0.9707$$

From equation (9-15), the parameter bias uncertainty at 0.5 years since calibration is projected to be

$$u(t) = u(0)\frac{\Phi^{-1}\left(\dfrac{1+R_0}{2}\right)}{\Phi^{-1}\left(\dfrac{1+R_t}{2}\right)} = 1.5\,\mu V \times \frac{\Phi^{-1}\left(\dfrac{1+0.98}{2}\right)}{\Phi^{-1}\left(\dfrac{1+0.9707}{2}\right)} = 1.5\,\mu V \times \frac{\Phi^{-1}(0.99)}{\Phi^{-1}(0.985)}$$

$$= 1.5\,\mu V \times \frac{2.3263}{2.1796} = 1.5\,\mu V \times 1.0673 = 1.60\,\mu V.$$

APPENDIX A – TERMS AND DEFINITIONS

It has been a goal of the authors to use consistent terminology throughout this document, even though the terms and definitions employed are designed to be understood across a broad technology base. Where appropriate, terms and definitions have been taken from internationally recognized standards and guidelines in the fields of testing and calibration.

Term	Definition
a priori value	A value assumed before measurements are taken.
Acceleration Error	Under steady-state conditions, it is the maximum difference, at any measurand value within the specified range, between output readings taken with and without the application of specified constant acceleration along specified axes. Acceleration error can also result from dynamic conditions encountered in vibration and shock environments.
Acceleration Sensitivity	See Acceleration Error.
Accuracy	Closeness of agreement between a declared or measured value of a quantity and its true value.
	In terms of instruments and other measuring devices, accuracy is defined as the conformity of an indicated value to the true value or, alternatively, to the value of an accepted standard.
Adjusted Mean	The value of a parameter or error source obtained by applying a correction factor to a nominal or mean value.
Amplifier	A device that increases the strength or amplitude of a signal.
Analog to Digital Converter (ADC)	A device that converts an analog signal to a digital representation of finite length.
Analog Signal	A quantity or signal that is continuous in both amplitude and time.
Arithmetic Mean	The sum of a set of values divided by the number of values in the set.
Artifact	A physical object or substance with measurable attributes.
Asymmetric Distribution	A probability distribution in which deviations from the population mode value of one sign are more probable than deviations of the opposite sign. Asymmetric distributions have a non-zero coefficient of skewness.
Attenuation	Reduction of signal strength, intensity or value.
Attribute	A measurable characteristic, feature, or aspect of a device, object or substance.
Attribute Bias	A systematic deviation of an attribute's nominal or indicated value from its true value.
Average	See Arithmetic Mean.

Term	Definition
Average-Over-Period (AOP) Reliability	The in-tolerance probability for a parameter averaged over its calibration or test interval. The AOP measurement reliability is often used to represent the in-tolerance probability of a parameter for a measuring device whose usage demand is random over its test or calibration interval.
Bandwidth	The range of frequencies that a device is capable of generating, handling, or accommodating; usually the range in which the response is within 3 dB of the maximum response.
Beginning-of-Period (BOP) Reliability	The in-tolerance probability for an MTE attribute or parameter at the start of its calibration or test interval.
Between Sample Sigma	The standard deviation representing the variation of values obtained for different samples taken on a given quantity.
Bias	A systematic discrepancy between an indicated, assume or declared value of a quantity and the quantity's true value. See also Attribute Bias and Operator Bias.
Bias Offset	See Offset from Nominal.
Bias Uncertainty	The uncertainty in the bias of an attribute or error source quantified as the standard deviation of the bias probability distribution.
Bit	A single character, 0 or 1, in a binary numeral system (base 2). The bit is the smallest unit of storage currently used in computing.
Calibration	A process in which the value of an MTE attribute or parameter is compared to a corresponding value of a measurement reference, resulting in (1) the determination that the parameter or attribute value is within its associated specification or tolerance limits, (2) a documented correction of the parameter or attribute value, or (3) a physical adjustment of the parameter or attribute value.
Calibration Interval	The scheduled interval of time between successive calibrations of one or more MTE parameter or attribute.
Characteristic	A distinguishing trait, feature or quality.
Combined Error	The error comprised of a combination of two or more error sources.
Combined Uncertainty	The uncertainty in a combined error.
Common Mode Rejection (CMR)	The common mode rejection ratio is often expressed in dB using the following relationship: CMR = 20 log(CMRR).
Common Mode Rejection Ratio (CMRR)	CMRR describes the ability of a differential amplifier to reject interfering signals common to both positive and negative input terminals, and to amplify only the difference between the inputs. Normally defined as the ratio of the signal gain to the ratio of normal mode voltage to the common mode voltage. CMMR = Gain/(NMV/CMV).

Term	Definition
Common Mode Voltage (CMV)	A voltage which is common to both input terminals of a differential device.
Compensation	Provision for a supplemental device, circuit, or special materials to counteract known sources of error.
Component Uncertainty	The product of the sensitivity coefficient and the standard uncertainty for an error source.
Computation Error	The error in a quantity obtained by computation. Computation error can result from machine round-off of values obtained by iteration or from the use of regression models. Sometimes applied to errors in tabulated physical constants.
Computed Mean Value	The average value of a sample of measurements.
Confidence Level	The probability that a set of tolerance or containment limits will contain a given error.
Confidence Limits	Limits that bound errors for a given source with a specified confidence level.
Containment Limits	Limits that are specified to contain either an attribute or parameter value, an attribute or parameter bias, or other measurement errors.
Containment Probability	The probability that an attribute or parameter value or error in the measurement of this value lies within specified containment limits.
Correlation Analysis	An analysis that determines the extent to which two random variables influence one another. Typically the analysis is based on ordered pairs of values. In the context of measurement uncertainty analysis, the random variables are error sources or error components.
Correlation Coefficient	A measure of the extent to which two errors are linearly related. A function of the covariance between the two errors. Correlation coefficients range from minus one to plus one.
Counts	The total number of divisions into which a given measurement range is divided. For example, a 5-1/2 digit voltmeter has +/- 199,999 or 399,999 total counts. The weight of a count is given by Count Weight = Total Range/Total Counts.
Covariance	The expected value of the product of the deviations of two random variables from their respective means. The covariance of two independent random variables is zero.
Coverage Factor	A factor used to express an error limit or expanded uncertainty as a multiple of the standard uncertainty.
Cross-correlation	The correlation between two errors for two different components of a multivariate analysis.
Cumulative Distribution Function	A mathematical function whose values $F(x)$ are the probabilities that a random variable assumes a value less than or equal to x. Synonymous with Distribution Function.

Term	Definition
Damping	The progressive reduction or suppression of the oscillation of a system.
Dead Band	The range through which the input varies without initiating a response (or indication) from the measuring device.
Degrees of Freedom	A statistical quantity that is related to the amount of information available about an uncertainty estimate. The degrees of freedom signifies how "good" the estimate is and serves as a useful statistic in determining appropriate coverage factors and computing confidence limits and other decision variables.
Deviation from Nominal	The difference between an attribute's or parameter's measured or true value and its nominal value.
Digital to Analog Converter (DAC)	A device for converting a digital (usually binary) code to a continuous, analog output.
Digital Signal	A quantity or signal that is represented as a series of discrete coded values.
Direct Measurements	Measurements in which a measuring parameter or attribute X directly measures the value of a subject parameter or attribute Y (i.e., X measures Y). In direct measurements, the value of the quantity of interest is obtained directly by measurement and is not determined by computing its value from the measurement of other variables or quantities.
Display Resolution	The smallest distinguishable difference between indications of a displayed value.
Distribution Function	See Cumulative Distribution Function.
Distribution Variance	The mean square dispersion of a random variable about its mean value. See also Variance.
Drift	A change in output over a period of time that is unrelated to input. Can be due to aging, temperature effects, sensor contamination, etc.
Dynamic Characteristics	Those characteristics of a measuring device that relate its response to variations of the physical input with time.
Dynamic Range	The range of physical input signals that can be converted to output signals by a measuring device.
Effective Degrees of Freedom	The degrees of freedom for Type B uncertainty estimates or a combined uncertainty estimate.
End-of-Period (EOP) Reliability	The in-tolerance probability for an MTE attribute or parameter at the end of its calibration or test interval.
Equipment Parameter	A specified aspect, feature or performance characteristic of a measuring device or artifact. Synonymous with MTE attribute or parameter.

Term	Definition
Error	The arithmetic difference between a measured or indicated value and the true value.
Error Component	The total error in a measured or assumed value of a component variable in a multivariate measurement. For example, in the determination of the volume of a right circular cylinder, there are two error components: the error in the length measurement and the error in the diameter measurement.
Error Distribution	A probability distribution that describes the relative frequency of occurrence of values of a measurement error.
Error Equation	An expression that defines the combined error in the value of a quantity in terms of all relevant process or component errors.
Error Limits	Bounding values that are expected to contain the error from a given source with some specified level of probability or confidence.
Error Model	See Error Equation.
Error Source	A parameter, variable or constant that can contribute error to the determination of the value of a quantity.
Error Source Coefficient	See Sensitivity Coefficient.
Error Source Correlation	See Correlation Analysis.
Error Source Uncertainty	The uncertainty in a given error source.
Estimated True Value	The value of a quantity obtained by Bayesian analysis.
Excitation	An external power supply required by measuring devices to convert a physical input to an electrical output. Typically, a well-regulated dc voltage or current.
Expanded Uncertainty	A multiple of the standard uncertainty reflecting either a specified confidence level or coverage factor.
False Accept Risk (FAR)	The probability that a measuring equipment attribute or parameter, accepted by conformance testing, will be out-or-tolerance. See NASA-HNBK-8739.19-4 for alternative definitions and applications.
Filter	A device that limits the signal bandwidth to reduce noise and other errors associated with sampling.
Frequency Response	The change with frequency of the output/input amplitude ratio (and of phase difference between output and input), for a sinusoidally varying input applied to a measuring device within a stated range of input frequencies.
Full Scale Input (FSI)	The arithmetic difference between the specified upper and lower input limits of a sensor, transducer or other measuring device.
Full Scale Output (FSO)	The arithmetic difference between the specified upper and lower output limits of a sensor, transducer or other measuring device.

Term	Definition
Gain	The ratio of the output signal to the input signal of an amplifier.
Gain Error	The degree to which gain varies from the ideal or target gain, specified in percent of reading.
Guardband	A supplement specification limit used to reduce the risk of falsely accepting a nonconforming or out-of-compliance MTE parameter.
Heuristic Estimate	An estimate resulting from accumulated experience and/or technical knowledge concerning the uncertainty of an error source.
Histogram	See Sample Histogram.
Hysteresis	The lagging of an effect behind its cause, as when the change in magnetism of a body lags behind changes in an applied magnetic field.
Hysteresis Error	The maximum separation due to hysteresis between upscale-going and downscale-going indications of a measured value taken after transients have decayed.
Independent Error Sources	Error sources that are statistically independent. See Statistical Independence.
Instrument	A device for measuring or producing the value of an observable quantity.
In-tolerance	In conformance with specified tolerance limits.
In-tolerance Probability	The probability that an MTE attribute or parameter value or the error in the value is contained within its specified tolerance limits at the time of measurement.
Kurtosis	A measure of the "peakedness" of a distribution. For example, normal distributions have a peakedness value of three.
Least Significant Bit (LSB)	The smallest analog signal value that can be represented with an n-bit code. LSB is defined as $A/2^n$, where A is the amplitude of the analog signal.
Level of Confidence	See Confidence Level.
Linearity	A characteristic that describes how a device's output over its range differs from a specified linear response.
Mean Deviation	The difference between a sample mean value and a nominal value.
Mean Square Error	See Variance.
Mean Value	*Sample Mean*: The average value of a measurement sample. *Population Mean*: The expectation value for measurements sampled from a population.
Mean Value Correction	The correction or adjustment of the computed mean value for an offset due to parameter bias and/or environmental factors.
Measurand	The particular quantity subject to measurement. (Taken from Annex B, Section B.2.9 of the GUM)

Term	Definition
Measurement Error	The difference between the measured value of a quantity and its true value.
Measurement Process Errors	Errors resulting from the measurement process (e.g., measurement reference bias, repeatability, resolution error, operator bias, environmental factors, etc).
Measurement Process Uncertainty	The uncertainty in a measurement process error. The standard deviation of the probability distribution of a measurement process error.
Measurement Reference	See Reference Standard.
Measurement Reliability	The probability that an MTE attribute or parameter is in conformance with performance specifications. At the measuring device or instrument level, it is the probablity that all attributes or parameters are in conformance or in-tolerance.
Measurement Uncertainty	The lack of knowledge of the sign and magnitude of measurement error.
Measurement Units	The units, such as volts, millivolts, etc., in which a measurement or measurement error is expressed.
Measuring Device	See Measuring and Test Equipment.
Measuring and Test Equipment (MTE)	A system or device used to measure the value of a quantity or test for conformance to specifications.
Measuring Parameter	The characteristic or feature of a measuring device that is used to obtain information that quantifies the value of the subject or unit-under-test parameter.
Median Value	(1) The value that divides an ordered sample of data in two equal portions. (2) The value for which the distribution function of a random variable is equal to one-half.
Mode Value	The value of a parameter most often encounter or measured. Sometimes synonymous with the nominal value or design value of a parameter.
Module Error Sources	Sources of error that accompany the conversion of module input to module output.
Module Input Uncertainty	The uncertainty in a module's input error expressed as the uncertainty in the output of the preceding module.
Module Output Equation	The equation that expresses the output from a module in terms of its input. The equation is characterized by parameters that represent the physical processes that participate in the conversion of module input to module output.
Module Output Uncertainty	The combined uncertainty in the output of a given module of a measurement system.

Term	Definition
Multiplexer	A multi-channel device designed to accept input signals from a number of sensors or measuring equipment and share downstream signal conditioning components.
Multivariate Measurements	Measurements in which the value of a subject parameter is a computed quantity based on measurements of two or more attributes or parameters.
Noise	Signals originating from sources other than those intended to be measured. Noise may arise from several sources, can be random or periodic, and often varies in intensity.
Nominal Value	The designated or published value of an artifact, attribute or parameter. It may also sometimes refer to the distribution mode value of an artifact, attribute or parameter.
Nonlinearity	See Linearity.
Normal Mode Voltage	The potential difference that exists between pairs of power (or signal) conductors.
Offset	A non-zero output of a device for a zero input.
Operating Conditions	The environmental conditions, such as pressure, temperature and humidity ranges that the measuring device is rated to operate.
Operator Bias	The systematic error due to the perception or influence of a human operator or other agency.
Output Device	See Readout Device.
Parameter	A characteristic of a device, process or function. See also Equipment Parameter.
Parameter Bias	A systematic deviation of a parameter's nominal or indicated value from its true value.
Population	The total set of possible values for a random variable.
Population Mean	The expectation value of a random variable described by a probability distribution.
Precision	The number of places past the decimal point in which the value of a quantity can be expressed. Although higher precision does not necessarily mean higher accuracy, the lack of precision in a measurement is a source of measurement error.
Probability	The likelihood of the occurrence of a specific event or value from a population of events or values.
Probability Density Function (pdf)	A mathematical function that describes the relative frequency of occurrence of the values of a random variable.
Quantization	The sub-division of the range of a reading into a finite number of steps, not necessary equal, each of which is assigned a value. Particularly applicable to analog to digital and digital to analog conversion processes.

Term	Definition
Quantization Error	Error due to the granularity of resolution in quantizing a sampled signal. Contained within +/- 1/2 LSB (least significant bit) limits.
Random Error	See Repeatability.
Range	An interval of values for which specified tolerances apply. In a calibration or test procedure, a setting or designation for the measurement of a set of specific points.
Rated Output (RO)	See Full Scale Output.
Readout Device	A device that converts a signal to a series of numbers on a digital display, the position of a pointer on a meter scale, tracing on recorder paper or graphic display on a screen.
Reference Standard	An artifact used as a measurement reference whose value and uncertainty have been determined by calibration and documented.
Reliability Model	A mathematical function relating the in-tolerance probability of one or more MTE attributes or parameters and the time between calibration. Used to project uncertainty growth over time.
Repeatability	The error that manifests itself in the variation of the results of successive measurements of a quantity carried out under the same measurement conditions and procedure during a measurement session. Often referred to as Random Error.
Reproducibility	The closeness of the agreement between the results of measurements of the value of a quantity carried out under different measurement conditions. The different conditions may include: principle of measurement, method of measurement, observer, measuring instrument(s), reference standard, location, conditions of use, time.
Resolution	The smallest discernible value indicated by a measuring device.
Resolution Error	The error due to the finiteness of the precision of a measurement.
Response Time	The time required for a sensor output to change from its previous state to a final settled value.
Sample	A collection of values drawn from a population from which inferences about the population are made.
Sample Histogram	A bar chart showing the relative frequency of occurrence of sampled values.
Sample Mean	The arithmetic average of sampled values.
Sample Size	The number of values that comprise a sample.
Sensitivity	The ratio between a change in the electrical output signal to a small change in the physical input of a sensor or transducer. The derivative of the transfer function with respect to the physical input.

Term	Definition
Sensitivity Coefficient	A coefficient that weights the contribution of an error source to a combined error.
Sensor	Any of various devices designed to detect, measure or record physical phenomena.
Settling Time	The time interval between the application of an input and the time when the output is within an acceptable band of the final steady-state value.
Signal Conditioner	A device that provides amplification, filtering, impedance transformation, linearization, analog to digital conversion, digital to analog conversion, excitation or other signal modification.
Skewness	A measure of the asymmetry of a probability distribution. A symmetric distribution has zero skewness.
Span	See Dynamic Range.
Specification	A numerical value or range of values that bound the performance of an MTE parameter or attribute.
Stability	The ability of a measuring device to give constant output for a constant input over a period of time.
Standard Deviation	The square root of the variance of a sample or population of values. A quantity that represents the spread of values about a mean value. In statistics, the second moment of a distribution.
Standard Uncertainty	The standard deviation of an error distribution.
Static Performance Characteristic	An indication of how the measuring equipment or device responds to a steady-state input at one particular time.
Statistical Independence	A property of two or more random variables such that their joint probability density function is the product of their individual probability density functions. Two error sources are statistically independent if one does not exert an influence on the other or if both are not consistently influenced by a common agency.
Stress Response Error	The error or bias in a parameter value induced by response to applied stress.
Student's t-statistic	Typically expressed as $t_{\alpha,\nu}$, it denotes the value for which the distribution function for a t-distribution with ν degrees of freedom is equal to $1 - \alpha$. A multiplier used to express an error limit or expanded uncertainty as a multiple of the standard uncertainty.
Subject Parameter	An attribute or quantity whose value we seek to obtain from a measurement or set of measurements.
Symmetric Distribution	A probability distribution of random variables that are equally likely to be found above or below a mean value.
System Equation	A mathematical expression that defines the value of a quantity in terms of its constituent variables or components.

Term	Definition
System Module	An intermediate stage of a system that transforms an input quantity into an output quantity according to a module output equation.
System Output Uncertainty	The total uncertainty in the output of a measurement system.
t Distribution	A symmetric, continuous distribution characterized by the degrees of freedom parameter. Used to compute confidence limits for normally distributed variables whose estimated standard deviation is based on a finite degrees of freedom. Also referred to as the Student's t-distribution.
Temperature Coefficient	A quantitative measure of the effects of a variation in operating temperature on a device's zero offset and sensitivity.
Temperature Effects	The effect of temperature on the sensitivity and zero output of a measuring device.
Thermal Drift	The change in output of a measuring device per degree of temperature change, given all other operating conditions are held constant.
Thermal Sensitivity Shift	The variation in the sensitivity of a measuring device as a function of temperature.
Thermal Transient Response	A change in the output from a measuring device generated by temperature change.
Thermal Zero Shift	The shift in the zero output of a measuring device due to change in temperature.
Threshold	The smallest change in the physical input that will result in a measurable change in transducer output.
Time Constant	The time required to complete 63.2% of the total rise or decay after a step change of input. It is derived from the exponential response $e^{-t/\tau}$ where t is time and τ is the time constant.
Tolerance Limits	Typically, engineering tolerances that define the maximum and minimum values for a product to work correctly. These tolerances bound a region that contains a certain proportion of the total population with a specified probability or confidence.
Total Module Uncertainty	See Module Output Uncertainty.
Total Uncertainty	The standard deviation of the probability distribution of the total combined error in the value of a quantity obtained by measurement.
Total System Uncertainty	See System Output Uncertainty.
Transducer	A device that converts an input signal from one form into an output signal of another form.
Transfer Function	A mathematical equation that shows the functional relationship between the physical input signal and the electrical output signal.

Term	Definition
Transient Response	The response of a measuring device to a step-change in the physical input. See also Response Time and Time Constant.
Transverse Sensitivity	An output caused by motion, which is not in the same axis that the device is designed to measure. Defined in terms of output for cross-axis input along the orthogonal axes.
True Value	The value that would be obtained by a perfect measurement. True values are by nature indeterminate.
Type A Estimates	Uncertainty estimates obtained by statistical analysis of a sample of data.
Type B Estimates	Uncertainty estimates obtained by heuristic means in the absence of a sample of data.
Uncertainty	See Standard Uncertainty.
Uncertainty Component	The uncertainty in an error component.
Uncertainty in the Mean Value	The standard deviation of the distribution of mean values obtained from multiple sample sets for a given measured quantity. Estimated by the standard deviation of a single sample set divided by the square root of the sample size.
Uncertainty Growth	The increase in the uncertainty in the value or bias of an MTE parameter or attribute over the time elapsed since measurement.
Variance	(1) Population: The expectation value for the square of the difference between the value of a variable and the population mean. (2) Sample: A measure of the spread of a sample equal to the sum of the squared observed deviations from the sample mean divided by the degrees of freedom for the sample. Also referred to as the mean square error.
Vibration Sensitivity	The maximum change in output, at any physical input value within the specified range, when vibration levels of specified amplitude and range of frequencies are applied to a transducer or other measuring device along specified axes.
Warm-up Time	The time it takes a circuit to stabilize after the application of power.
Within Sample Sigma	An indicator of the variation within samples.
Zero Balance	See Offset.
Zero Drift	See Zero Shift.
Zero Offset	See Offset.
Zero Shift	A change in the output of a measuring device, for a zero input, over a specified period of time.

APPENDIX B – PROBABILITY DISTRIBUTIONS

A probability distribution is a relationship between the value of a variable and its probability of occurrence. Such distributions may be characterized by different degrees of spreading or may even exhibit different shapes. Probability distributions are usually expressed as a mathematical function $f(\varepsilon)$ called the **probability density function**, or pdf.

Axiom 1 tells us that measurement errors are random variables that follow probability distributions. For certain kinds of error, such as repeatability or random error, the validity of this assertion is easily seen. Conversely, for other kinds of error, such as parameter bias and operator bias, the validity of this assertion may not be so readily apparent.

It is important to bear in mind, however, that, while a particular error may have a systematic value that persists from measurement to measurement, it nevertheless comes from some distribution of like errors that possess a probability of occurrence. Consequently, whether a particular error is random or systematic, it can be regarded as coming from a distribution of errors that can be described statistically.

Once the probability distribution for a measurement error has been characterized, the uncertainty in this error can be computed. The uncertainty for a given error source, ε, is equal to the square root of the distribution variance.

$$u_\varepsilon = \sqrt{\operatorname{var}(\varepsilon)} \qquad \text{(B-1)}$$

where

$$\operatorname{var}(\varepsilon) = \int_{-\infty}^{\infty} f(\varepsilon)\,(\varepsilon - \mu)^2\, d\varepsilon \qquad \text{(B-2)}$$

For symmetric error distributions, the population mean μ is taken to be zero. In these cases, equation (B-2) reduces to

$$\operatorname{var}(\varepsilon) = \int_{-\infty}^{\infty} f(\varepsilon)\,\varepsilon^2\, d\varepsilon \qquad \text{(B-3)}$$

This appendix describes probability distributions that can be used to characterize measurement errors. Once the probability distribution for a measurement error has been characterized, the uncertainty in this error is computed as the square root of the distribution variance. Because the Uniform distribution is often incorrectly selected as a simple means of obtaining an uncertainty estimate, Section B.12 is included to discuss its proper application.

B.1   Normal Distribution

When obtaining a Type A uncertainty estimate, we compute a standard deviation from a sample of values. For example, the uncertainty due to repeatability is estimated by computing the standard deviation for a sample of repeated measurements of a given value. The sample standard deviation is an estimate of the standard deviation for the population from which the sample was drawn. Except in rare cases, we assume that this population follows the normal distribution.

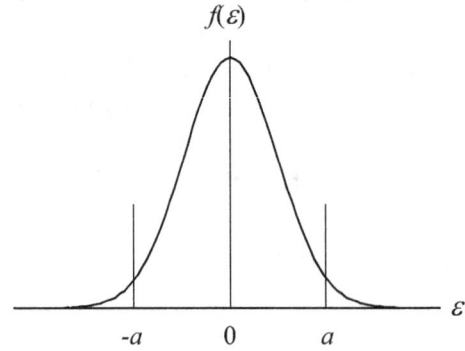

Figure B-1. Normal Distribution

Why do we assume a normal distribution? The primary reason is because this is the distribution that either represents or approximates what we frequently see in the physical universe. It can be derived from the laws of physics for such phenomena as the diffusion of gases and is applicable to instrument parameters subject to random stresses of usage and handling. It is also often applicable to equipment parameters emerging from manufacturing processes.

In addition, the normal distribution is often assumed applicable for a total error composed of constituent errors. This assumption results from the *central limit theorem*, which demonstrates that, even though the individual constituent errors may not be normally distributed, the combined error is approximately so.

The probability density function for the normal distribution is given in equation (B-4). The population mean is equal to zero and the variable σ is the population standard deviation.

$$f(\varepsilon) = \frac{1}{\sqrt{2\pi}\sigma} e^{-\varepsilon^2 / 2\sigma^2} \tag{B-4}$$

In applying the normal distribution, an uncertainty estimate is obtained from containment limits and a containment probability.

For example, if $\pm a$ represents the known containment limits and p represents the associated containment probability, then an uncertainty estimate can be obtained from equation (B-5).

$$u_\varepsilon = \frac{a}{\Phi^{-1}\left(\dfrac{1+p}{2}\right)} \tag{B-5}$$

The inverse normal distribution function, $\Phi^{-1}()$, can be found in statistics texts and in most spreadsheet programs. If only a single containment limit is applicable, such as with single-sided tolerance limits, the appropriate expression is given in equation (B-6).

$$u_\varepsilon = \frac{a}{\Phi^{-1}(p)} \tag{B-6}$$

Note: The use of the normal distribution is appropriate in cases where the above considerations apply and the limits and probability are at least approximately known. The extent to which this knowledge is approximate determines the degrees of freedom of the uncertainty estimate. The degrees of freedom and the uncertainty estimate can be used in conjunction with the Student's t distribution to compute confidence limits. The Student's t distribution is discussed in Section B.10.

B.2   Lognormal Distribution

The lognormal distribution can often be used to estimate the uncertainty in equipment parameter bias in cases where the tolerance limits are asymmetric. This distribution is also used in cases where a physical limit is present that lies close enough to the nominal or mode value to skew the probability density function in such a way that the normal distribution is not applicable.

A typical right-handed lognormal distribution with physical limit q, mode $M = 0$ and two-sided, asymmetric tolerance limits - a_1 and a_2 is shown in Figure B-2.

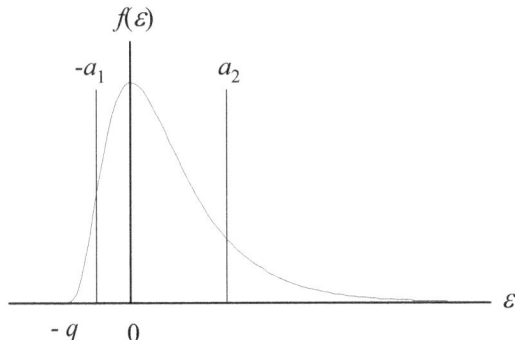

Figure B-2. Right-handed Lognormal Distribution

The probability density function for the lognormal distribution is given in equation (B-7). The variable q is a physical limit for ε, the variable m is the population median and the variable λ is the shape parameter. The quantities m, q and λ are obtained by numerical iteration, given containment limits and an associated containment probability.

$$f(\varepsilon) = \frac{1}{\sqrt{2\pi}\lambda|\varepsilon + q|}\exp\left\{-\left[\ln\left(\frac{\varepsilon+q}{m+q}\right)\right]^2 \Big/ 2\lambda^2\right\} \qquad (B\text{-}7)$$

The lognormal distribution statistics are defined in equations (B-8) through (B-11).

Mode: $M = 0$ $\qquad\qquad\qquad\qquad\qquad\qquad\qquad\qquad$ (B-8)

Median: $m = q\left(e^{\lambda^2} - 1\right)$ $\qquad\qquad\qquad\qquad\qquad\qquad$ (B-9)

Mean: $\mu = (m+q)e^{\lambda^2/2} - q$ $\qquad\qquad\qquad\qquad\qquad$ (B-10)

150

Variance: $\quad \mathrm{var}(\varepsilon) = \sigma^2 = (m+q)^2 e^{\lambda^2}(e^{\lambda^2}-1)$ $\qquad$ (B-11)

The uncertainty is the square root of the variance.

$$u_\varepsilon = |m+q| e^{\lambda^2/2}\sqrt{e^{\lambda^2}-1} \qquad \text{(B-12)}$$

B.3   Exponential Distribution

Sometimes cases are encountered where there exists a definable upper or lower physical limit to errors along with a single-sided upper or lower tolerance limit. If the physical limit and the mode value are equal, then the lognormal distribution suffers from a mathematical discontinuity that makes it inappropriate as the distribution of choice. To handle such cases, the exponential distribution is employed. A plot of a right-handed exponential distribution is shown in Figure B-3 where the mode $M = 0$ is less than the tolerance limit a.

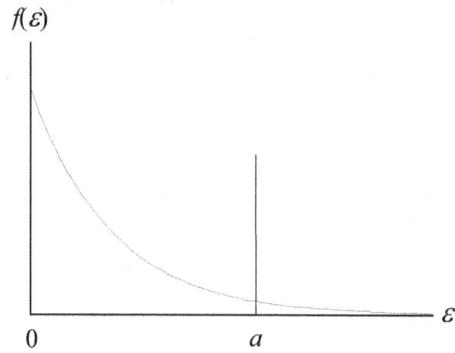

Figure B-3. Right-handed Exponential Distribution

The probability density function for the exponential distribution is given in equation (B-13).

$$f(\varepsilon) = \lambda e^{-\lambda|\varepsilon|} \qquad \text{(B-13)}$$

The absolute value for ε is used to accommodate cases where the tolerance limit a is less than zero, as depicted in Figure B-4.

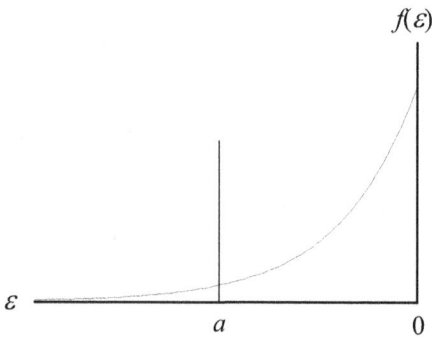

Figure B-4. Left-Handed Exponential Distribution

Employing the probability density function for a right-handed exponential distribution, the variance is

$$\text{var}(\varepsilon) = \int_0^\infty \lambda e^{-\lambda\varepsilon} (\varepsilon - \mu)^2 \, d\varepsilon \qquad\qquad \text{(B-14)}$$

The distribution mean is computed to be

$$\mu = \int_0^\infty \lambda e^{-\lambda\varepsilon} \varepsilon \, d\varepsilon = \lambda \int_0^\infty e^{-\lambda\varepsilon} \varepsilon \, d\varepsilon$$

$$= -\lambda \left[\frac{e^{-\lambda\varepsilon}}{\lambda^2} (\lambda\varepsilon + 1) \right]_0^\infty = -\lambda \left[\frac{e^{-\infty}}{\lambda^2} (\infty + 1) - \frac{e^{-0}}{\lambda^2} (0 + 1) \right] \qquad\qquad \text{(B-15)}$$

$$= -\lambda \left[0 - \frac{1}{\lambda^2} \right] = \frac{1}{\lambda}$$

Note: The mean value for a left-handed exponential distribution is $\mu = -\dfrac{1}{\lambda}$.

Substituting equation (B-15) into equation (B-14), the variance of the right-handed distribution is computed to be

$$\text{var}(\varepsilon) = \int_0^\infty \lambda e^{-\lambda\varepsilon} \left(\varepsilon - \frac{1}{\lambda} \right)^2 d\varepsilon = \lambda \int_0^\infty e^{-\lambda\varepsilon} \left(\varepsilon^2 - 2\frac{\varepsilon}{\lambda} + \frac{1}{\lambda^2} \right) d\varepsilon$$

$$= \lambda \int_0^\infty e^{-\lambda\varepsilon} \varepsilon^2 d\varepsilon - 2 \int_0^\infty e^{-\lambda\varepsilon} \varepsilon \, d\varepsilon + \frac{1}{\lambda} \int_0^\infty e^{-\lambda\varepsilon} d\varepsilon$$

$$= \lambda \left[\frac{\varepsilon^2 e^{-\lambda\varepsilon}}{-\lambda} \bigg|_0^\infty + \frac{2}{\lambda} \int_0^\infty e^{-\lambda\varepsilon} \varepsilon \, d\varepsilon \right] + 2 \left[\frac{e^{-\lambda\varepsilon}}{\lambda^2} (\lambda\varepsilon + 1) \right]_0^\infty - \frac{1}{\lambda} \left[\frac{e^{-\lambda\varepsilon}}{\lambda} \right]_0^\infty$$

$$\qquad\qquad \text{(B-16)}$$

$$= -\left[\varepsilon^2 e^{-\lambda\varepsilon} \right]_0^\infty - 2 \left[\frac{e^{-\lambda\varepsilon}}{\lambda^2} (\lambda\varepsilon + 1) \right]_0^\infty + 2 \left[\frac{e^{-\lambda\varepsilon}}{\lambda^2} (\lambda\varepsilon + 1) \right]_0^\infty - \frac{1}{\lambda} \left[\frac{e^{-\lambda\varepsilon}}{\lambda} \right]_0^\infty$$

$$= -\left[\infty^2 e^{-\infty} - 0 e^{-0} \right] - \frac{1}{\lambda} \left[\frac{e^{-\infty}}{\lambda} - \frac{e^{-0}}{\lambda} \right]$$

$$= -\frac{1}{\lambda} \left[0 - \frac{1}{\lambda} \right] = \frac{1}{\lambda^2}$$

Note: The variance of a left-handed exponential distribution is equivalent to the variance of a right-handed exponential distribution.

The uncertainty estimate for the exponential distribution is obtained by taking the square root of the variance.

$$u_\varepsilon = \frac{1}{\lambda} \qquad \text{(B-17)}$$

B.4   Quadratic Distribution

The quadratic distribution, shown in Figure B-5, is continuous between minimum bounding limits, does not exhibit unrealistic linear behavior and satisfies the need for a central tendency.

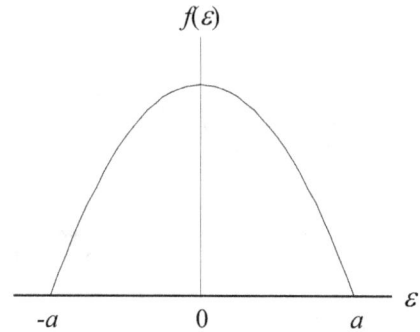

Figure B-5. Quadratic Distribution

For containment limits ±L and associated containment probability p, the minimum bounding limits ±a are obtained from equation (B-18).

$$a = \frac{L}{2p}\left(1 + 2\cos\left[\frac{1}{3}\arccos(1-2p)\right]\right), \quad L \le a \qquad \text{(B-18)}$$

The quadratic distribution is defined by the probability density function given in equation (B-19).

$$f(\varepsilon) = \begin{cases} \dfrac{3}{4a}\left[1-(\varepsilon/a)^2\right], & -a \le \varepsilon \le a \\ 0, & \text{otherwise} \end{cases} \qquad \text{(B-19)}$$

The quadratic distribution variance is

$$\begin{aligned}
\text{var}(\varepsilon) &= \frac{3}{4a}\int_{-a}^{a}\left[1-(\varepsilon/a)^2\right]\varepsilon^2 d\varepsilon = \frac{3}{4a}\int_{-a}^{a}\left[\varepsilon^2 - \frac{\varepsilon^4}{a^2}\right]d\varepsilon = \frac{3}{4a}\left[\frac{\varepsilon^3}{3} - \frac{\varepsilon^5}{5a^2}\right]_{-a}^{a} \\
&= \frac{3}{4a}\left[\frac{2a^3}{3} - \frac{2a^5}{5a^2}\right] = \frac{3a^2}{2}\left[\frac{1}{3} - \frac{1}{5}\right] = \frac{3a^2}{2}\left[\frac{5}{15} - \frac{3}{15}\right] \qquad \text{(B-20)} \\
&= \frac{3a^2}{2}\frac{2}{15} = \frac{a^2}{5}
\end{aligned}$$

The uncertainty is the square root of the variance.

$$u_\varepsilon = \frac{a}{\sqrt{5}} \qquad \text{(B-21)}$$

B.5 Cosine Distribution

While the quadratic distribution is continuous within minimum bounding limits, it is discontinuous at the limits. And, even though the quadratic distribution has wider applicability than either the triangular or uniform distribution, this feature nevertheless diminishes its physical validity. As shown in Figure B-6, the cosine distribution overcomes this shortcoming, exhibits a central tendency and can be determined from bounding limits.

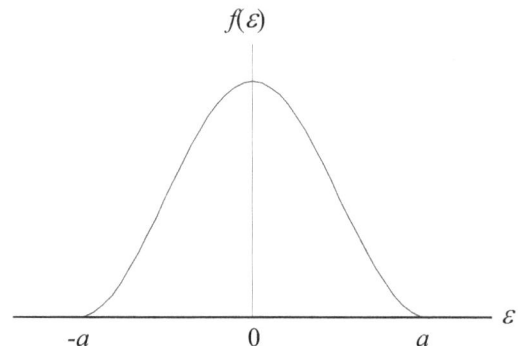

Figure B-6. Cosine Distribution

Given containment limits $\pm L$ and associated containment probability p, the minimum bounding limits $\pm a$ are computed by solving equation (B-22) using a numerical iterative method.

$$\frac{a}{\pi}\sin\left(\pi L/a\right) - ap + L = 0, \quad L \le a \qquad \text{(B-22)}$$

The iterative algorithm is given in equations (B-23) through (B-25).

$$a_i = a_{i-1} - F_i / F_i' \qquad \text{(B-23)}$$

$$F_i = \frac{a_i}{\pi}\sin(\pi L/a_i) - a_i p + L \qquad \text{(B-24)}$$

$$F_i' = \frac{1}{\pi}\sin(\pi L/a_{i-1}) - \frac{L}{a}\cos(\pi L/a_{i-1}) - p \qquad \text{(B-25)}$$

where a_i is the value obtained at the ith iteration.

The probability density function for the cosine distribution is given in equation (B-26).

$$f(\varepsilon) = \begin{cases} \dfrac{1}{2a}\left[1 + \cos\left(\pi\varepsilon/a\right)\right], & -a \le \varepsilon \le a \\ 0, & \text{otherwise} \end{cases} \qquad \text{(B-26)}$$

154

The cosine distribution variance is

$$\text{var}(\varepsilon) = \frac{1}{2a} \int_{-a}^{a} \left[1 + \cos\left(\pi\varepsilon/a\right)\right] \varepsilon^2 d\varepsilon = \frac{1}{2a} \int_{-a}^{a} \left[\varepsilon^2 + \varepsilon^2 \cos\left(\pi\varepsilon/a\right)\right] d\varepsilon$$

$$= \frac{1}{2a} \left[\frac{\varepsilon^3}{3} + \frac{2a^2 \varepsilon \cos\left(\pi\varepsilon/a\right)}{\pi^2} + \frac{\left(\frac{\pi}{a}\right)^2 \varepsilon^2 - 2}{\left(\frac{\pi}{a}\right)^3} \sin\left(\pi\varepsilon/a\right) \right]_{-a}^{a} \tag{B-27}$$

$$= \frac{1}{2a}\left[\frac{2a^3}{3} + \frac{2a^3 \cos(\pi)}{\pi^2} + \frac{2a^3 \cos(-\pi)}{\pi^2}\right] = \frac{1}{2a}\left[\frac{2a^3}{3} - \frac{4a^3}{\pi^2}\right]$$

$$= \frac{a^2}{3} - \frac{2a^2}{\pi^2} = \frac{a^2}{3}\left(1 - \frac{6}{\pi^2}\right)$$

The uncertainty is the square root of the variance.

$$u_\varepsilon = \frac{a}{\sqrt{3}} \sqrt{1 - \frac{6}{\pi^2}} \tag{B-28}$$

Note: The value of u_ε for the cosine distribution translates to roughly 63% of the value obtained using the uniform distribution.

B.6   U-shaped Distribution

The U-shaped distribution shown in Figure B-7 applies to sinusoidal RF signals incident on a load. Another application for this distribution would be environmental temperature control in a laboratory or test chamber.

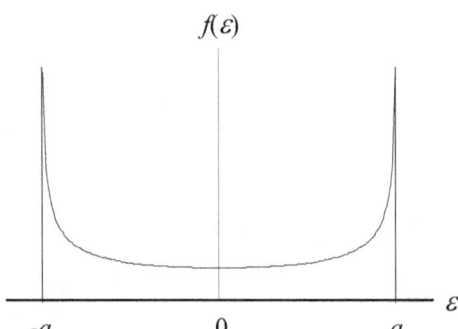

Figure B-7. U-Shaped Distribution

If containment limits $\pm L$ and containment probability p are known, the minimum bounding limits $\pm a$ can be computed from equation (B-29).

$$a = \frac{L}{\sin(\pi p/2)}, \quad L \leq a \tag{B-29}$$

The probability density function for the U-shaped distribution is given in equation (B-30).

$$f(\varepsilon) = \begin{cases} \dfrac{1}{\pi\sqrt{a^2 - \varepsilon^2}}, & -a \leq \varepsilon \leq a \\ 0, & \text{otherwise} \end{cases} \tag{B-30}$$

The U-shaped distribution variance is

$$
\begin{aligned}
\text{var}(\varepsilon) &= \frac{1}{\pi} \int_{-a}^{a} \frac{\varepsilon^2}{\sqrt{a^2 - \varepsilon^2}} d\varepsilon \\
&= \frac{1}{\pi}\left[-\frac{\varepsilon}{2}\sqrt{a^2 - \varepsilon^2} + \frac{a^2}{2}\sin^{-1}(\varepsilon/a) \right]_{-a}^{a} \\
&= \frac{1}{\pi}\left[\frac{a^2}{2}\sin^{-1}(1) - \frac{a^2}{2}\sin^{-1}(-1) \right] \\
&= \frac{1}{\pi}\left[\frac{a^2}{2}\left(\frac{\pi}{2}\right) - \frac{a^2}{2}\left(-\frac{\pi}{2}\right) \right] = \frac{1}{\pi}\left[\frac{a^2\pi}{2} \right] = \frac{a^2}{2}
\end{aligned}
\tag{B-31}
$$

The uncertainty is the square root of the variance.

$$u_\varepsilon = \frac{a}{\sqrt{2}} \tag{B-32}$$

Note: The value of u_ε for the U-shaped distribution translates to roughly 122% of the value obtained using the uniform distribution.

B.7   Uniform (Rectangular) Distribution

The uniform distribution has minimum bounding limits and an equal probability of obtaining a value within these limits. There are two types of uniform distribution.

- The "round-off" uniform distribution

- The "truncation" uniform distribution

B.7.1   Round-off Uniform Distribution

The round-off uniform distribution describes errors that fall within symmetric minimum bounding limits $\pm a$ centered at zero, as shown in Figure B-8. The probability of lying between the minimum bounding limits is constant and the probability of lying outside of them is zero.

The probability density function for the round-off uniform distribution is

$$f(\varepsilon) = \begin{cases} \dfrac{1}{2a}, & -a \le \varepsilon \le a \\[2mm] 0, & \text{otherwise} \end{cases} \qquad \text{(B-33)}$$

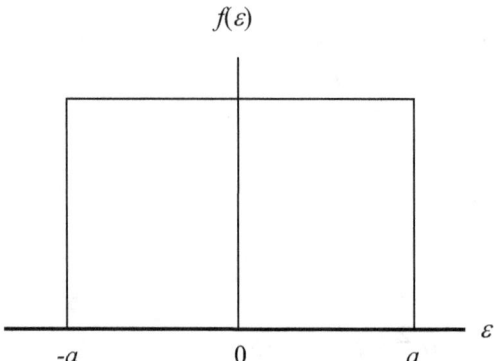

Figure B-8. Round-off Uniform Distribution

The round-off uniform distribution variance is

$$\text{var}(\varepsilon) = \int_{-a}^{a} \frac{1}{2a}\varepsilon^2 d\varepsilon = \frac{1}{2a}\int_{-a}^{a}\varepsilon^2 d\varepsilon$$

$$= \frac{1}{a}\int_{0}^{a}\varepsilon^2 d\varepsilon = \frac{1}{a}\frac{\varepsilon^3}{3}\Bigg|_0^a = \frac{1}{a}\frac{a^3}{3} = \frac{a^2}{3} \qquad \text{(B-34)}$$

The uncertainty is computed by taking the square root of the variance.

$$u_\varepsilon = \frac{a}{\sqrt{3}} \qquad \text{(B-35)}$$

B.7.2   Truncation Uniform Distribution

The "truncation" uniform distribution describes errors that are distributed between the limits 0 and a, as shown in Figure B-9. The probability density function for the truncation uniform distribution is given in equation (B-36).

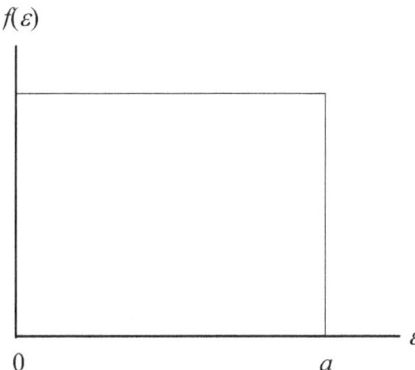

Figure B-9. Truncation Uniform Distribution

$$f(\varepsilon) = \begin{cases} \dfrac{1}{a}, & 0 \le \varepsilon \le a \\ 0, & \text{otherwise} \end{cases} \tag{B-36}$$

The truncation uniform distribution variance is computed to be

$$\begin{aligned} \text{var}(\varepsilon) &= \int_0^a \frac{1}{a}\varepsilon^2 d\varepsilon = \frac{1}{a}\int_0^a \varepsilon^2 d\varepsilon \\ &= \frac{1}{a}\frac{\varepsilon^3}{3}\Big|_0^a = \frac{1}{a}\frac{a^3}{3} = \frac{a^2}{3} \end{aligned} \tag{B-37}$$

The uncertainty is computed by taking the square root of the variance.

$$u_\varepsilon = \frac{a}{\sqrt{3}} \tag{B-38}$$

B.8   Triangular Distribution

The triangular distribution, shown in Figure B-10, is the simplest distribution possible for use in cases where there are minimum containment limits and there is a central tendency for values of the error.

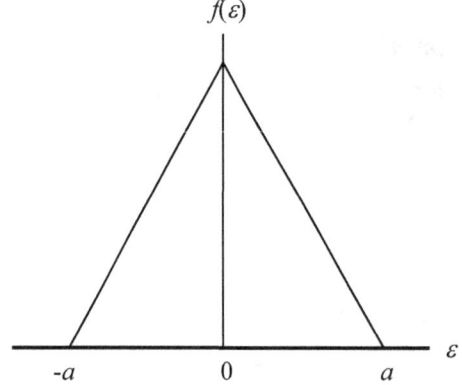

$f(\varepsilon)$

$-a$ 0 a ε

Figure B-10. Triangular Distribution

The triangular distribution sometimes applies to parameter values immediately following test or calibration and to the sum of two uniformly distributed errors that have the same mean value and minimum bounding limits.

Apart from representing post-test distributions under certain restricted conditions, the triangular distribution has limited applicability. While it does not suffer from the constant probability criterion of the uniform distribution, it nevertheless displays abrupt transitions at the bounding limits and at the zero point, which are physically unrealistic in most instances.

The probability density function for the triangular distribution is given in equation (B-39).

$$f(\varepsilon) = \begin{cases} (a+\varepsilon)/a^2, & -a \le \varepsilon \le 0 \\ (a-\varepsilon)/a^2, & 0 \le \varepsilon \le a \\ 0, & \text{otherwise.} \end{cases} \tag{B-39}$$

The triangular distribution variance is computed to be

$$\begin{aligned} \text{var}(\varepsilon) &= \frac{1}{a^2}\int_{-a}^{0}(a+\varepsilon)\varepsilon^2 d\varepsilon + \frac{1}{a^2}\int_{0}^{a}(a-\varepsilon)\varepsilon^2 d\varepsilon \\ &= \frac{1}{a^2}\int_{0}^{a}(a-\varepsilon)\varepsilon^2 d\varepsilon = \frac{2}{a}\int_{0}^{a}\varepsilon^2 d\varepsilon - \frac{2}{a^2}\int_{0}^{a}\varepsilon^3 d\varepsilon \\ &= \left[\frac{2}{a}\frac{\varepsilon^3}{3} - \frac{2}{a^2}\frac{\varepsilon^4}{4}\right]_{0}^{a} = \frac{2}{a}\frac{a^3}{3} - \frac{2}{a^2}\frac{a^4}{4} \\ &= \frac{2a^2}{3} - \frac{a^2}{2} = \frac{4a^2}{6} - \frac{3a^2}{6} = \frac{a^2}{6} \end{aligned} \tag{B-40}$$

The uncertainty is computed by taking the square root of the variance.

159

$$u_\varepsilon = \sqrt{\mathrm{var}(\varepsilon)} = \sqrt{\frac{a^2}{6}} = \frac{a}{\sqrt{6}} \qquad (\text{B-41})$$

B.9 Trapezoidal Distribution

If two errors ε_x and ε_y are uniformly distributed with bounding values of $\pm a$ and $\pm b$, where $b \geq a$, then their sum

$$\varepsilon = \varepsilon_x + \varepsilon_y$$

follows a trapezoidal distribution with discontinuities at $\pm c = \pm(b - a)$ and $\pm d = \pm(b + a)$, as shown in Figure (B-11).

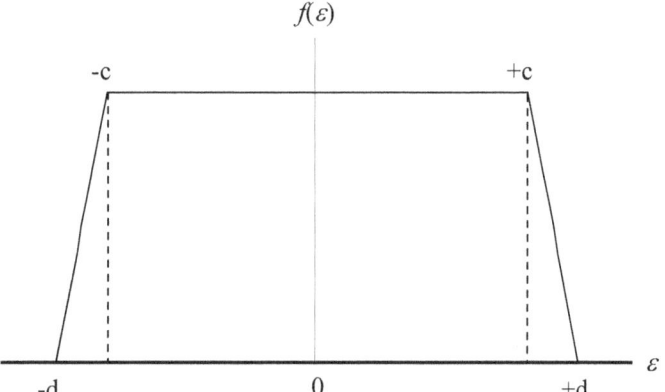

Figure B-11. Trapezoidal Distribution

The probability of obtaining a value of ε is uniform between the limits $\pm c$, declining linearly to zero at the minimum bounding limits $\pm d$.

When applying the trapezoidal distribution, it may be difficult to establish the minimum bounding limits $\pm d$. One approach would be to specify tolerance limits $\pm L$, with an associate in-tolerance probability p, and uniform probability limits $\pm c$. This information could then be used to solve for d. If this approach is used, there are two possible cases.

Case 1:

If $L \leq c$, then the in-tolerance probability is

$$p = \frac{2L}{d + c} \qquad (\text{B-42})$$

and

$$d = \frac{2L}{p} - c \qquad (\text{B-43})$$

Case 2:

If $c < L < d$, then the in-tolerance probability is

160

$$p = 1 - \frac{(d-L)^2}{d^2 + c^2} \tag{B-44}$$

and d is obtained by solving equation (B-45).

$$d = \frac{L}{p}\left\{1 \pm \sqrt{(1-p)\left[1 - p(c/L)^2\right]}\right\} \tag{B-45}$$

Note that, when $p = 1$, $d = L$.

The probability density function for the trapezoidal distribution is

$$f(\varepsilon) = \begin{cases} \dfrac{1}{d^2 - c^2}(d+\varepsilon), & -d \le \varepsilon \le -c \\[2mm] \dfrac{1}{d+c}, & -c \le \varepsilon \le c \\[2mm] \dfrac{1}{d^2 - c^2}(d-\varepsilon), & c \le \varepsilon \le d \\[2mm] 0, & \text{otherwise} \end{cases} \tag{B-46}$$

The trapezoidal distribution variance is

$$\begin{aligned}
\mathrm{var}(\varepsilon) &= \frac{1}{d^2 - c^2}\int_{-d}^{-c}(d+\varepsilon)\varepsilon^2 d\varepsilon + \frac{1}{d+c}\int_{-c}^{c}\varepsilon^2 d\varepsilon + \frac{1}{d^2 - c^2}\int_{c}^{d}(d-\varepsilon)\varepsilon^2 d\varepsilon \\[2mm]
&= \frac{2}{d^2 - c^2}\int_{c}^{d}(d-\varepsilon)\varepsilon^2 d\varepsilon + \frac{2}{d+c}\int_{0}^{c}\varepsilon^2 d\varepsilon \\[2mm]
&= \frac{2}{d^2 - c^2}\int_{c}^{d}\left(d\varepsilon^2 - \varepsilon^3\right)d\varepsilon + \frac{2}{d+c}\int_{0}^{c}\varepsilon^2 d\varepsilon \\[2mm]
&= \frac{2}{d^2 - c^2}\left[\frac{d\varepsilon^3}{3} - \frac{\varepsilon^4}{4}\right]_{c}^{d} + \frac{2}{d+c}\left[\frac{\varepsilon^3}{3}\right]_{0}^{c} \\[2mm]
&= \frac{2}{d^2 - c^2}\left[\frac{d^4}{3} - \frac{d^4}{4} - \frac{dc^3}{3} + \frac{c^4}{4}\right] + \frac{2}{d+c}\frac{c^3}{3} \\[2mm]
&= \frac{1}{d^2 - c^2}\left[\frac{4d^4}{6} - \frac{3d^4}{6} - \frac{4dc^3}{6} + \frac{3c^4}{6}\right] + \frac{4}{6}\frac{(d-c)c^3}{d^2 - c^2} \\[2mm]
&= \frac{d^4 - 4dc^3 + 3c^4 + 4dc^3 - 4c^4}{6(d^2 - c^2)} = \frac{d^4 - c^4}{6(d^2 - c^2)}
\end{aligned} \tag{B-47}$$

161

The uncertainty is the square root of the variance.

$$u_\varepsilon = \sqrt{\frac{d^4 - c^4}{6\left(d^2 - c^2\right)}}$$ (B-48)

B.10 Student's t Distribution

If the underlying distribution is normal, and a Type A estimate and degrees of freedom are available, confidence limits for measurement errors may be obtained using the Student's t distribution.

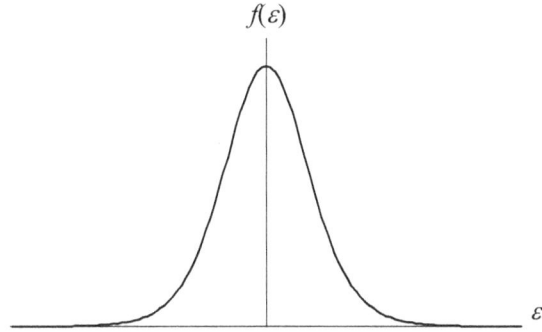

$f(\varepsilon)$

ε

Figure B-12. Student's t Distribution

The probability density function for the Student's t distribution is given in equation (B-49). The variable ν is the degrees of freedom and the parameter $\Gamma(\cdot)$ is the gamma function.

$$f\left(\varepsilon\right) = \frac{\Gamma\left(\dfrac{\nu+1}{2}\right)}{\sqrt{\pi\nu}\,\Gamma\left(\dfrac{\nu}{2}\right)}\left(1 + \frac{\varepsilon^2}{\nu}\right)^{-\left(\frac{\nu+1}{2}\right)}$$ (B-49)

The degrees of freedom quantifies the amount of knowledge used in estimating uncertainty. For Type A estimates the degrees of freedom is simply the sample size, n, minus one, as shown in equation (B-50).

$$\nu = n - 1$$ (B-50)

The knowledge used in estimating uncertainty is incomplete if containment limits $\pm a$ for the Student's t distribution are approximate and the containment probability p is estimated from recollected experience (i.e., Type B). Therefore, the degrees of freedom associated with a Type B estimate is not infinite.

If the degrees of freedom are finite but unknown, the uncertainty estimate cannot be rigorously used to develop confidence limits, perform statistical tests or make decisions. This limitation has often precluded the use of Type B estimates as statistical quantities and has led to the misguided practice of using fixed coverage factors.

162

Fortunately, the GUM provides an expression for obtaining the approximate degrees of freedom for Type B estimates. However, the expression involves the use of the variance in the uncertainty estimate, and a method for obtaining this variance has been lacking until recently.[70]

The procedure is to first estimate the uncertainty for the normal distribution and then estimate the degrees of freedom from the following expression

$$v_B \cong \frac{1}{2}\left(\frac{\text{var}(u_\varepsilon)}{u_\varepsilon^2}\right)^{-1} \cong \frac{3\varphi^2 L^2}{2\varphi^2(\Delta L)^2 + \pi a^2 e^{\varphi^2}(\Delta p)^2} \qquad \text{(B-51)}$$

where the variables ΔL and Δp represent "give or take" values for the containment limits and containment probability, respectively, and

$$\varphi = \Phi^{-1}\left(\frac{1+p}{2}\right) \qquad \text{(B-52)}$$

Once the degrees of freedom has been obtained, the Type B estimate can be combined with other estimates and the degrees of freedom for the combined uncertainty can be determined using the Welch-Satterthwaite formula. If the underlying distribution for the combined estimate is normal, the Student's t distribution can be used to develop confidence limits and perform statistical tests.

For confidence or containment limits $\pm L$ and corresponding degrees of freedom, the uncertainty can be estimated from

$$u_\varepsilon = \frac{L}{t_{\alpha/2,v}} \qquad \text{(B-53)}$$

where $t_{\alpha/2,v}$ is the Student's t statistic, $\alpha = 1 - p$, and p is the containment probability or confidence level. The Student's t statistic for a given set of $\alpha/2$ and v values can be obtained from published tables.[71]

B.11   The Utility Distribution

In some cases, one might expect the probability of finding a measurement error to be essentially uniform over a range of values, tapering off gradually to zero at the distribution limits. The utility distribution, shown in Figure B-13, represents this behavior. This distribution gets its name because of its application to building utility functions in cost analysis applications.

---

[70] Type B degrees of freedom estimation is discussed in Appendix D.
[71] See for example, *CRC Standard Mathematical Tables*, 28th Edition, CRC Press Inc., 2000.

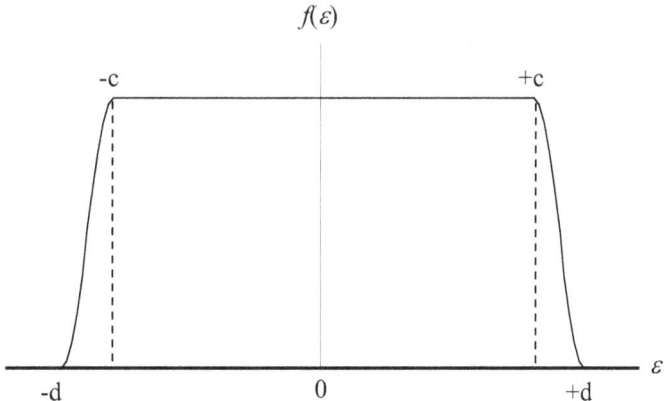

Figure B-13. Utility Distribution

The utility distribution is very similar to the trapezoidal distribution, including the difficulty establishing the limits $\pm c$ and $\pm d$. As with the trapezoidal distribution, the approach is to specify tolerance limits $\pm L$, with an associate in-tolerance probability p, and uniform probability limits. This information could then be used to solve for d. There are two cases.

Case 1:

If $L \leq c$, then the in-tolerance probability is

$$p = \frac{2L}{d+c}$$
(B-54)

and

$$d = \frac{2L}{p} - c$$
(B-55)

Case 2:

If $c < L < d$, then the in-tolerance probability is

$$p = \frac{1}{d+c}\left\{L + c + \frac{d-c}{\pi}\sin\left[\frac{\pi(L-c)}{d-c}\right]\right\}$$
(B-56)

The probability density function for the utility distribution is

$$f(\varepsilon) = \begin{cases} \dfrac{1}{d+c}, & |\varepsilon| \leq c \\ \dfrac{1}{d+c}\cos^2\left[\dfrac{(|\varepsilon|-c)\pi}{2(d-c)}\right], & c \leq |\varepsilon| \leq d \\ 0, & |\varepsilon| \geq d \end{cases}$$
(B-57)

The utility distribution variance is

164

$$\text{var}\left(\varepsilon\right) = \frac{1}{d+c}\left[\int_{-c}^{c}\varepsilon^2 d\varepsilon + \int_{-d}^{-c}\cos^2\left[\frac{\pi(\varepsilon+c)}{2(d-c)}\right]\varepsilon^2 d\varepsilon + \int_{c}^{d}\cos^2\left[\frac{\pi(\varepsilon-c)}{2(d-c)}\right]\varepsilon^2 d\varepsilon\right]$$

$$= \frac{2}{d+c}\left[\int_{0}^{c}\varepsilon^2 d\varepsilon + \int_{c}^{d}\cos^2\left[\frac{\pi(\varepsilon-c)}{2(d-c)}\right]\varepsilon^2 d\varepsilon\right]$$

(B-58)

Substituting the variable

$$\zeta = \frac{\pi\left(\varepsilon-c\right)}{2\left(d-c\right)}$$

into equation (B-58), the distribution variance becomes

$$\text{var}\left(\varepsilon\right) = \frac{2}{d+c}\int_{0}^{c}\varepsilon^2 d\varepsilon + \frac{4(d-c)}{\pi(d+c)}\int_{0}^{\pi/2}\cos^2\zeta\left[\frac{2}{\pi}(d-c)\zeta + c\right]^2 d\zeta$$

$$= \frac{2}{d+c}\left[\frac{\varepsilon^3}{3}\right]_{0}^{c} + \frac{4(d-c)}{\pi(d+c)}\int_{0}^{\pi/2}\left[\frac{4}{\pi^2}(d-c)^2\zeta^2 + \frac{4}{\pi}c(d-c)\zeta + c^2\right]\cos^2\zeta d\zeta$$ (B-59)

$$= \frac{2c^3}{3(d+c)} + I_1 + I_2 + I_3$$

where I_1, I_2 and I_3 are defined in equations (B-60) through (B-62).

$$I_1 = \frac{16(d-c)^3}{\pi^3(d+c)}\int_{0}^{\pi/2}\zeta^2\cos^2\zeta d\zeta$$

$$= \frac{16(d-c)^3}{\pi^3(d+c)}\left[\frac{\zeta^3}{6} + \frac{1}{4}\zeta^2\sin 2\zeta - \frac{1}{8}\sin 2\zeta + \frac{1}{4}\zeta\cos 2\zeta\right]_{0}^{\pi/2}$$

$$= \frac{16(d-c)^3}{\pi^3(d+c)}\left[\frac{\pi^3}{48} + \frac{\pi^2}{16}\sin\pi - \frac{1}{8}\sin\pi + \frac{\pi}{8}\cos\pi\right]$$ (B-60)

$$= \frac{16(d-c)^3}{\pi^3(d+c)}\left[\frac{\pi^3}{48} - \frac{\pi}{8}\right] = \frac{2(d-c)^3}{\pi^2(d+c)}\left[\frac{\pi^2}{6} - 1\right]$$

165

$$I_2 = \frac{4(d-c)}{\pi(d+c)} \frac{4}{\pi} c(d-c) \int\limits_{0}^{\pi/2} \zeta \cos^2 \zeta d\zeta$$

$$= \frac{16c(d-c)^2}{\pi^2(d+c)} \int\limits_{0}^{\pi/2} \zeta \cos^2 \zeta d\zeta$$

$$= \frac{16c(d-c)^2}{\pi^2(d+c)} \left[\frac{\zeta^2}{4} + \frac{\zeta \sin 2\zeta}{4} + \frac{\cos 2\zeta}{8} \right]_0^{\pi/2}$$

$$= \frac{16c(d-c)^2}{\pi^2(d+c)} \left[\frac{\pi^2}{16} + \frac{\pi \sin \pi}{8} + \frac{\cos \pi}{8} - \frac{\cos 0}{8} \right]$$

$$= \frac{16c(d-c)^2}{\pi^2(d+c)} \left[\frac{\pi^2}{16} - \frac{1}{4} \right] = \frac{c(d-c)^2}{\pi^2(d+c)} \left(\pi^2 - 4 \right)$$

(B-61)

$$I_3 = \frac{4c^2(d-c)}{\pi(d+c)} \int\limits_{0}^{\pi/2} \cos^2 \zeta d\zeta$$

$$= \frac{4c^2(d-c)}{\pi(d+c)} \left[\frac{\zeta}{2} + \frac{\sin 2\zeta}{4} \right]_0^{\pi/2}$$

$$= \frac{4c^2(d-c)}{\pi(d+c)} \left[\frac{\pi}{4} + \frac{\sin \pi}{4} - \frac{\sin 0}{4} \right]$$

$$= \frac{4c^2(d-c)}{\pi(d+c)} \left[\frac{\pi}{4} \right] = \frac{c^2(d-c)}{d+c}$$

(B-62)

Substituting equations (B-60) through (B-62) into equation (B-59), the utility distribution variance is computed to be

$$\text{var}(\varepsilon) = I_1 + I_2 + I_3 + \frac{2c^3}{3(d+c)}$$

$$= \frac{2(d-c)^3}{\pi^2(d+c)} \left(\frac{\pi^2}{6} - 1 \right) + \frac{c(d-c)^2}{\pi^2(d+c)} \left(\pi^2 - 4 \right) + \frac{c^2(d-c)}{d+c} + \frac{2c^3}{3(d+c)}$$

$$= \frac{2(d-c)^3 \left(\frac{1}{6} - \frac{1}{\pi^2} \right) + c(d-c)^2 \left(1 - \frac{4}{\pi^2} \right) + c^2 d - \frac{1}{3}c^3}{d+c}$$

(B-63)

$$= \frac{2(d^3 - 3cd^2 + 3c^2 d - c^3) \left(\frac{1}{6} - \frac{1}{\pi^2} \right) + c(d^2 - 2cd + c^2) \left(1 - \frac{4}{\pi^2} \right) + c^2 d - \frac{1}{3}c^3}{d+c}$$

Rearranging equation (B-63),

$$\text{var}(\varepsilon) = \frac{1}{d+c}\left[\begin{array}{l} \dfrac{d^3}{3} - cd^2 + c^2 d - \dfrac{c^3}{3} - \dfrac{2}{\pi^2}(d^3 - 3cd^2 + 3c^2 d - c^3) \\[2mm] + cd^2 - 2c^2 d + c^3 - \dfrac{4}{\pi^2}(cd^2 - 2c^2 d + c^3) + c^2 d - \dfrac{c^3}{3} \end{array} \right]$$

$$= \frac{1}{d+c}\left[\frac{d^3}{3} + \frac{c^3}{3} - \frac{2}{\pi^2}\left(d^3 - cd^2 - c^2 d + c^3\right) \right] \qquad \text{(B-64)}$$

$$= \frac{1}{d+c}\left[\frac{d^3 + c^3}{3} - \frac{2(d-c)^2(d+c)}{\pi^2} \right]$$

$$= \frac{d^3 + c^3}{3(d+c)} - \frac{2(d-c)^2}{\pi^2}$$

The uncertainty is the square root of the variance.

$$u_\varepsilon = \sqrt{\frac{d^3 + c^3}{3(d+c)} - \frac{2(d-c)^2}{\pi^2}} \qquad \text{(B-65)}$$

B.12 Applying the Uniform Distribution

Unfortunately, the uniform distribution is often incorrectly selected for Type B estimates because an uncertainty is simply computed by dividing containment limits by the square root of three. Advocates of the adhoc use of the uniform distribution have asserted that this practice is recommended in the GUM.

Basic selection criteria are provided herein, including specific cases where the uniform distribution is applicable. The two common fallacies for universal or adhoc application of the uniform distribution are also dispelled.

B.12.1 Criteria for Selecting the Uniform Distribution

The use of the uniform distribution is appropriate under a limited set of conditions. These conditions are summarized by three criteria.

1. The minimum bounding limits must be known for the distribution. This is the minimum limits criterion.

2. There must be a 100% probability of finding values between these limits. This is the 100% containment criterion.

3. There must be equal probability of obtaining values between the minimum bounding limits. This is the equal probability criterion.

Minimum Limits Criterion
It is vital that the limits established for the uniform distribution are the minimum bounding limits. For instance, if the limits $\pm a$ bound the error distribution, then so do the limits $\pm 2a$, $\pm 3a$, and so on. Since the uncertainty estimate for the uniform distribution is obtained by dividing the

bounding limit by the square root of three, using a value for the limit that is not the minimum bounding value will obviously result in an invalid uncertainty estimate.

This alone makes the application of the uniform distribution questionable in estimating bias uncertainty from such quantities as tolerance limits. It may be that out-of-tolerances have never been observed for a particular parameter (100% containment), but it is unknown whether the tolerances are minimum bounding limits.

A difficulty often encountered when attempting to apply minimum bounding limits is that such limits can rarely be established on physical grounds. This is especially true when using tolerance limits for a given MTE parameter.

Some years ago, a study was conducted involving a voltage reference that showed that values for one parameter were normally distributed with a standard deviation that was approximately 1/10 of the tolerance limit. With 10-sigma limits, it is unlikely that any out-of-tolerances would be observed. However, if the uniform distribution were used to estimate the bias uncertainty for this item, based on tolerance limits, the uncertainty estimate would be nearly six times larger than would be appropriate. Some might claim that this is acceptable, since the estimate can be considered a conservative one. That may be. However, it is also a unrealistic estimate.

100% Containment Criterion

By definition, the establishment of minimum bounding limits implies the establishment of 100% containment. It should be said however, that an uncertainty estimate may still be obtained for the uniform distribution if a containment probability less that 100% is applied. For instance, suppose the containment limits are given as $\pm L$ and the containment probability is stated as being equal to some value p between zero and one. Then, if the uniform probability criterion is met, the minimum bounding limits of the distribution are given by

$$ a = \frac{L}{p}, \quad L \le a $$

If the equal probability criterion is not met, however, the uniform distribution would not be applicable, and we should turn to other distributions.

Equal Probability Criterion

As discussed above, establishing minimum bounding limits can be a challenging prospect. It is harder still to find real-world measurement error distributions that demonstrate an equal probability of occurrence between two limits and zero probability of occurrence outside these limits. Except in very limited instances, such as those discussed in Section B.12.2, assuming equal probability is not physically realistic.

B.12.2 Cases where the Uniform Distribution is Applicable

Digital Resolution Uncertainty

We sometimes need to estimate the uncertainty due to the resolution of a digital readout. For instance, a three-digit readout might indicate 12.015 V. If the device employs the standard round-off practice, we know that the displayed number is derived from a sensed value that lies between 12.0145 V and 12.0155 V. We also can assert to a very high degree of validity that the value has an equal probability of lying anywhere between these two numbers. In this case, the use of the uniform distribution is appropriate, and the resolution uncertainty is

$$u_{\varepsilon_{res}} = \frac{0.0005 \text{ V}}{\sqrt{3}} = 0.00029 \text{ V} .$$

RF Phase Angle

RF power incident on a load may be delivered to the load with a phase angle θ between $-\theta$ and θ. Unless there is a compelling reason to believe otherwise, the probability of occurrence between these limits is equal (i.e., uniform). This yields a phase angle uncertainty estimate of

$$u_\theta = \frac{\pi}{\sqrt{3}} \cong 1.814 .$$

> **Note**: Given the above, if we assume that the amplitude of the signal is sinusoidal, the distribution for incident voltage is the U-shaped distribution.

Quantization Error

When an analog signal is digitized, the sampled signal points are quantized in multiples of a discrete step size. The potential drop (or lack of a potential drop) sensed across each element of an analog to digital converter (ADC) sensing network produces either a "1" or "0" to the converter. This response constitutes a "bit" in the binary code that represents the sampled value.

Even if no errors were present in sampling and sensing the input signal, errors would still be introduced by the discrete nature of the encoding process. Suppose, for example, that the full scale signal level (dynamic range) of the ADC is V_m volts. If n bits are used in the encoding process, then the voltage can be resolved into 2^n discrete steps, each of size $V_m/2^n$.

The containment limit associated with each step is one-half the value of the magnitude of the step. Consequently, the containment limits for quantization error are $\pm V_m/2^{n+1}$. The uncertainty due to quantization error is obtained from these containment limits and from the assumption that there is equal probability of occurrence between these limits.

$$u_{\varepsilon_{quant}} = \frac{V_m / 2^{n+1}}{\sqrt{3}} .$$

B.12.3 Incorrect Application of the Uniform Distribution

The indiscriminate use of the uniform distribution to obtain Type B uncertainty estimates is a practice that has been gaining ground over the past few years. The two main reasons for this are

1. Ease of use.

2. Recommended in the GUM.

Ease of Use

Applying the uniform distribution makes it easy to obtain an uncertainty estimate. If the minimum bounding limits of the distribution are known, the uncertainty estimate, u, is simply computed from dividing these limits by the square root of three.

It should be said that the "ease of use" advantage has been promoted by individuals who are ignorant of methods of obtaining uncertainty estimates for more appropriate distributions and by others who are simply looking for a quick solution.

In fairness to the latter group, they sometimes assert that the lack of specificity of information required to use other distributions makes for crude uncertainty estimates anyway, so why not get your crude estimate by intentionally using an inappropriate distribution?

Since the introduction of the GUM, methods have been developed to systematically and rigorously apply distributions that are physically realistic.

Recommended in the GUM
It has been asserted by some that the use of the uniform distribution is recommended in the GUM. In fact, most of the methodology of the GUM is based on the assumption that the underlying error distribution is normal. For clarification on this issue, the reader is referred to Section 4.3 of the GUM.

Another source of confusion is that some of the examples in the GUM apply the uniform distribution in situations that appear to be incompatible with its use. It is reasonable to suppose that much of this is due to the fact that rigorous Type B estimation methods and tools were not available at the time the GUM was published, and the uniform distribution was an "easy out."

The philosophy of indiscriminately using the uniform distribution to compute Type B uncertainty estimates undermines efforts to estimate uncertainties that can be used to perform statistical tests, evaluate measurement decision risks, manage calibration intervals, develop meaningful tolerances and compute viable confidence limits.

APPENDIX C – STATISTICAL SAMPLE ANALYSIS

In the real world, it is seldom practical or economical to obtain all possible values of a population. Instead, a data sample is drawn from a population of interest and sample statistics are used to make inferences about the characteristics of the population. Three commonly used data sampling formats (sampled values, sampled cells and sampled mean values) and their relevant statistics are discussed herein.

In taking samples of measurement data, we collect the results of some number of repeat measurements. For the sample statistics to be meaningful, we must ensure that each measurement is both **independent** and **representative**. Measurements are independent if, in measuring one value, we do not affect the measurement of another value. Measurements are representative if they are typical of the kind of measurements we are interested in obtaining.[72] We must also ensure that the data are sampled **randomly**. In this regard, we strive to collect the data "as it comes" without any screening that may skew the results.

When making repeat measurements, it is also important to include all sampled values, provided they are independent and representative – not just the ones that appeal to us. However, this does not mean that clearly anomalous values should be included. Methods for statistically identifying outliers from samples of measurement data are presented in Section C.4.

The normal distribution is often assumed to be the underlying distribution for randomly sampled data. However, this assumption may not apply to all measurement sampling scenarios. Section C.5 discusses normality testing to determine if sampled data can be assumed to be normally distributed.

A question that commonly arises when making repeat measurements is "what is considered to be a sufficient sample size?" Section C.6 addresses the effect of sample size on computed statistics and presents a method that can be used to determine if the size of a sample of data is sufficient for obtaining an estimated sample mean that differs from the true (population) mean by less than or equal to some specified amount.

C.1   Sampled Values

In this format, sampled values consist of individual repeat measurements. The data can be expressed as measured values or deviations from a nominal or specified value.

The sample **mean**, $\bar{x}$, is obtained by taking the average of the sampled values. The average value is computed by summing the sampled values and dividing them by the sample size, n.

$$\bar{x} = \frac{1}{n}\left(x_1 + x_2 + \ldots + x_n\right) = \frac{1}{n}\sum_{i=1}^{n} x_i \tag{C-1}$$

The sample **standard deviation** provides an estimate of how much the population is spread about the mean value. The sample standard deviation, s_x, is computed by taking the square root of the sum of the squares of sampled deviations from the mean divided by the sample size minus one.

---

[72] i.e., if they are obtained from the population of interest.

$$s_x = \sqrt{\frac{1}{n-1} \sum_{i=1}^{n} (x_i - \bar{x})^2}$$ (C-2)

The value n-1 is the degrees of freedom for the estimate, which signifies the number of independent pieces of information that go into computing the estimate. All other things being equal, the greater the degrees of freedom, the closer the sample estimate will be to its population counterpart. The degrees of freedom for an uncertainty estimate is useful for establishing confidence limits and other decision variables.

The standard deviation in the mean value, $s_{\bar{x}}$, is equal to the standard deviation s_x divided by the square root of the sample size.

$$s_{\bar{x}} = \frac{s_x}{\sqrt{n}}$$ (C-3)

C.1.1 Example 1 – AC Voltage Measurements

In this example, measurements are made to evaluate the repeatability of the AC voltage coming out of a wall socket. We will compute the mean value and standard deviation for the voltage data listed below.

Measurement	AC Voltage
1	115.5
2	116.0
3	116.5
4	114.3
5	115.3
6	117.1
7	115.2
8	116.2
9	115.2
10	115.5
11	116.0
12	115.8
13	115.5
14	116.5
15	117.2

The sample mean is computed to be

$$\bar{V} = \frac{1}{15} \left(\begin{array}{l} 115.5 + 116.0 + 116.5 + 114.3 + 115.3 + 117.1 + 115.2 + 116.2 \\ +115.2 + 115.5 + 116.0 + 115.8 + 115.5 + 116.5 + 117.2 \end{array} \right) V_{ac}$$

$$= \frac{1737.8 \; V_{ac}}{15} = 115.9 \; V_{ac}$$

and the differences between the measured values and the mean value are

172

$$V_1 - \bar{V} = 115.5 - 115.9 = -0.4 \text{ V}_{ac}$$
$$V_2 - \bar{V} = 116.0 - 115.9 = 0.1 \text{ V}_{ac}$$
$$V_3 - \bar{V} = 116.5 - 115.9 = 0.6 \text{ V}_{ac}$$
$$V_4 - \bar{V} = 114.3 - 115.9 = -1.6 \text{ V}_{ac}$$
$$V_5 - \bar{V} = 115.3 - 115.9 = -0.6 \text{ V}_{ac}$$
$$V_6 - \bar{V} = 117.1 - 115.9 = 1.2 \text{ V}_{ac}$$
$$V_7 - \bar{V} = 115.2 - 115.9 = -0.7 \text{ V}_{ac}$$
$$V_8 - \bar{V} = 116.2 - 115.9 = 0.3 \text{ V}_{ac}$$

$$V_9 - \bar{V} = 115.2 - 115.9 = -0.7 \text{ V}_{ac}$$
$$V_{10} - \bar{V} = 115.5 - 115.9 = -0.4 \text{ V}_{ac}$$
$$V_{11} - \bar{V} = 116.0 - 115.9 = 0.1 \text{ V}_{ac}$$
$$V_{12} - \bar{V} = 115.8 - 115.9 = -0.1 \text{ V}_{ac}$$
$$V_{13} - \bar{V} = 115.5 - 115.9 = -0.4 \text{ V}_{ac}$$
$$V_{14} - \bar{V} = 116.5 - 115.9 = 0.6 \text{ V}_{ac}$$
$$V_{15} - \bar{V} = 117.2 - 115.9 = 1.3 \text{ V}_{ac}$$

The standard deviation is

$$s_V = \sqrt{\frac{1}{14}\left[\begin{array}{l}(-0.4)^2 + (0.1)^2 + (0.6)^2 + (-1.6)^2 + (-0.6)^2 + (1.2)^2 + (-0.7)^2 + (0.3)^2 \\ + (-0.7)^2 + (-0.4)^2 + (0.1)^2 + (-0.1)^2 + (-0.4)^2 + (0.6)^2 + (1.3)^2\end{array}\right]} \text{ V}_{ac}$$

$$= \sqrt{\frac{8.35}{14}} \text{ V}_{ac} = 0.77 \text{ V}_{ac}$$

and the standard deviation of the mean value is

$$s_{\bar{V}} = \frac{0.77 \text{ V}_{ac}}{\sqrt{15}} = 0.20 \text{ V}_{ac} .$$

C.1.2 Example 2 – Temperature Measurements

In this example, a digital thermometer is calibrated in a temperature bath using a standard platinum resistance thermometer (SPRT) as the temperature reference. The bath temperature is set so that the SPRT reads 100.000 °C and the thermometer temperature is recorded. This procedure is repeated several times. We will compute the mean and standard deviation of the temperature data listed below.

Measurement	SPRT °C	Thermometer °C	Deviation °C
1	100.000	100.02	0.02
2	100.000	100.03	0.03
3	100.000	99.98	-0.02
4	100.000	100.02	0.02
5	100.000	100.03	0.03
6	100.000	100.02	0.02
7	100.000	99.99	-0.01

The sample mean is computed to be

$$\bar{T} = 100.00\,°\text{C} + \frac{1}{7}\left(0.02 + 0.03 - 0.02 + 0.02 + 0.02 + 0.03 - 0.01\right)°\text{C}$$

$$= 100.00\,°\text{C} + \frac{0.09\,°\text{C}}{7} = 100.00\,°\text{C} + 0.01\,°\text{C} = 100.01\,°\text{C}$$

and the differences between the measured values and the mean value are

$$T_1 - \bar{T} = 100.02 - 100.01 = 0.01\,°\text{C}$$
$$T_2 - \bar{T} = 100.03 - 100.01 = 0.02\,°\text{C}$$
$$T_3 - \bar{T} = 99.98 - 100.01 = -0.03\,°\text{C}$$
$$T_4 - \bar{T} = 100.02 - 100.01 = 0.01\,°\text{C}$$
$$T_5 - \bar{T} = 100.02 - 100.01 = 0.01\,°\text{C}$$
$$T_6 - \bar{T} = 100.03 - 100.01 = 0.02\,°\text{C}$$
$$T_7 - \bar{T} = 99.99 - 100.01 = -0.02\,°\text{C}$$

The standard deviation is

$$s_T = \sqrt{\frac{1}{6}\left[(0.01)^2 + (0.02)^2 + (-0.03)^2 + (0.01)^2 + (0.01)^2 + (0.02)^2 + (-0.02)^2\right]}\,°\text{C}$$

$$= \sqrt{\frac{0.0035}{6}}\,°\text{C} = 0.02\,°\text{C}$$

and the standard deviation of the mean value is

$$s_{\bar{T}} = \frac{0.02\,°\text{C}}{\sqrt{7}} = 0.008\,°\text{C}.$$

C.2   Sampled Cells

In this data sampling format, sample values consist of repeat measurements that have been observed one or more times. The data are comprised of measured values or deviations from nominal, along with the number of times that a value has been observed.

The sample mean, $\bar{x}$, is obtained by taking the average of the sampled cell values. The average value is computed by summing the sampled cell values and dividing them by the sample size, n.

$$\bar{x} = \frac{1}{n}\sum_{i=1}^{k} n_i x_i \tag{C-4}$$

where
    n_i =   sample size or number of observations of a given sampled value, x_i
    k =   number of sampled cells
and

$$n = \sum_{i=1}^{k} n_i . \tag{C-5}$$

The sample standard deviation, s_x, is computed by taking the square root of the sum of the squares of sampled deviations from the mean divided by the sample size minus one.

$$s_x = \sqrt{\frac{1}{n-1} \sum_{i=1}^{k} n_i \left(x_i - \bar{x}\right)^2} \qquad \text{(C-6)}$$

The standard deviation in the mean value is computed as

$$s_{\bar{x}} = \frac{s_x}{\sqrt{n}}. \qquad \text{(C-7)}$$

C.2.1 Example 1 – AC Voltage Measurements

In this example, we will use the same AC voltage measurement data used in C.1.1 arranged into sample cells.

Sample Cell	AC Voltage	Number Observed
1	115.5	3
2	116.0	2
3	116.5	2
4	114.3	1
5	115.3	1
6	117.1	1
7	115.2	2
8	116.2	1
9	115.8	1
10	117.2	1

The sample mean value is computed to be

$$\bar{V} = \frac{1}{15} \left(\begin{array}{l} 3 \times 115.5 + 2 \times 116.0 + 2 \times 116.5 + 114.3 + 115.3 + 117.1 + 2 \times 115.2 \\ + 116.2 + 115.8 + 117.2 \end{array} \right) V_{ac}$$

$$= \frac{1737.8 \ V_{ac}}{15} = 115.9 \ V_{ac}$$

and the differences between the sampled cell values and the mean value are

$$V_1 - \bar{V} = 115.5 - 115.9 = -0.4 \ V_{ac}$$
$$V_2 - \bar{V} = 116.0 - 115.9 = 0.1 \ V_{ac}$$
$$V_3 - \bar{V} = 116.5 - 115.9 = 0.6 \ V_{ac}$$
$$V_4 - \bar{V} = 114.3 - 115.9 = -1.6 \ V_{ac}$$
$$V_5 - \bar{V} = 115.3 - 115.9 = -0.6 \ V_{ac}$$

175

$$V_6 - \bar{V} = 117.1 - 115.9 = 1.2 \ V_{ac}$$
$$V_7 - \bar{V} = 115.2 - 115.9 = -0.7 \ V_{ac}$$
$$V_8 - \bar{V} = 116.2 - 115.9 = 0.3 \ V_{ac}$$
$$V_9 - \bar{V} = 115.8 - 115.9 = -0.1 \ V_{ac}$$
$$V_{10} - \bar{V} = 117.2 - 115.9 = 1.3 \ V_{ac}$$

The sample standard deviation is

$$s_V = \sqrt{\frac{1}{14}\left[\begin{array}{l} 3\times(-0.4)^2 + 2\times(0.1)^2 + 2\times(0.6)^2 + (-1.6)^2 + (-0.6)^2 + (1.2)^2 \\ +2\times(-0.7)^2 + (0.3)^2 + (-0.1)^2 + (1.3)^2 \end{array}\right]} \ V_{ac}$$

$$= \sqrt{\frac{8.35}{14}} \ V_{ac} = 0.77 \ V_{ac}$$

and the standard deviation of the mean value is

$$s_{\bar{V}} = \frac{0.77 \ V_{ac}}{\sqrt{15}} = 0.20 \ V_{ac} \ .$$

C.3   Sampled Mean Values

In this format, the data sample consists of mean values obtained from sets of repeat measurements. The data are comprised of mean values or mean deviations from nominal value, along with the standard deviation and sample size for each set of repeat measurements.

For illustration, assume that our sample consists of k mean values and that the ith mean value, $\bar{x}_i$, and standard deviation of the ith sample, s_i, have been computed via a spreadsheet or other program using the following equations

$$\bar{x}_i = \frac{1}{n_i}\sum_{j=1}^{n_i} x_{ij} \tag{C-8}$$

$$s_i = \sqrt{\frac{1}{n_i - 1}\sum_{j=1}^{n_i}(x_{ij} - \bar{x}_i)^2} \tag{C-9}$$

where

n_i	=	the ith sample size
$\bar{x}_i$	=	mean value for the ith sample (i.e., ith mean value)
s_i	=	standard deviation of ith sample
x_{ij}	=	the jth measurement of the ith sample

The overall mean value, $\bar{x}$, (i.e., of all measurements) is computed from equation (C-10).

176

$$\bar{x} = \frac{1}{n} \sum_{i=1}^{k} \sum_{j=1}^{n_i} x_{ij} = \frac{1}{n} \sum_{i=1}^{k} n_i \bar{x}_i \qquad (\text{C-10})$$

where

k = number of samples (i.e., number of mean values entered)

n = total number of measurements (i.e., cumulative of all sample sizes)

= $\sum_{i=1}^{k} n_i$

The standard deviation of the sampled mean values relative to the overall mean value is the *between sample sigma*, s_b, computed from equation (C-11).

$$s_b = \sqrt{\frac{1}{n-1} \sum_{i=1}^{k} n_i (\bar{x}_i - \bar{x})^2} \qquad (\text{C-11})$$

An indicator of the variation within samples is the *within sample sigma*, s_w, computed from equation (C-12).

$$s_w = \sqrt{\frac{1}{n-1} \sum_{i=1}^{k} (n_i - 1) s_i^2} \qquad (\text{C-12})$$

The standard deviation, s, of all x_i values is computed from equation (C-13).

$$\begin{aligned}
s &= \sqrt{\frac{1}{n-1} \sum_{i=1}^{k} \sum_{j=1}^{n_i} (x_{ij} - \bar{x})^2} \\
&= \sqrt{\frac{1}{n-1} \sum_{i=1}^{k} n_i (\bar{x}_i - \bar{x})^2 + \frac{1}{n-1} \sum_{i=1}^{k} (n_i - 1) s_i^2} \\
&= \sqrt{s_b^2 + s_w^2}
\end{aligned} \qquad (\text{C-13})$$

The standard deviation for the mean of the sample mean values is computed by taking the variance of $\bar{x}$.

$$\begin{aligned}
\text{var}(\bar{x}) &= \text{var}\left(\frac{1}{n} \sum_{i=1}^{k} n_i \bar{x}_i \right) \\
&= \frac{1}{n^2} \sum_{i=1}^{k} n_i^2 \, \text{var}(\bar{x}_i)
\end{aligned} \qquad (\text{C-14})$$

From equation (C-1),

$$\text{var}\left(\overline{x}_i\right) = \text{var}\left(\frac{1}{n_i} \sum_{j=1}^{n_i} x_{ij}\right)$$

$$= \frac{1}{n_i^2} \sum_{j=1}^{n_i} \text{var}\left(x_{ij}\right) \tag{C-15}$$

Since each x_{ij} is sampled from a population with a variance equal to σ_x^2, then $\text{var}(x_{ij}) = \sigma_x^2$ and equation (C-15) becomes

$$\text{var}\left(\overline{x}_i\right) = \frac{1}{n_i^2} \sum_{j=1}^{n_i} \text{var}\left(x_{ij}\right)$$

$$= \frac{1}{n_i^2} \sum_{j=1}^{n_i} \sigma_x^2 = \frac{n_i \sigma_x^2}{n_i^2} = \frac{\sigma_x^2}{n_i} \tag{C-16}$$

Substituting equation (C-16) into (C-14), gives

$$\text{var}\left(\overline{x}\right) = \frac{1}{n^2} \sum_{i=1}^{k} n_i \sigma_x^2 = \frac{n \sigma_x^2}{n^2} = \frac{\sigma_x^2}{n} = \sigma_{\overline{x}}^2 \tag{C-17}$$

where $\sigma_{\overline{x}}$ is the standard deviation of the mean value population.

The population standard deviation σ_x is estimated with the sample standard deviation s computed from equation (C-13). Similarly, the standard deviation of the mean of the mean values can be estimated from equation (C-18).

$$s_{\overline{x}} = \frac{s}{\sqrt{n}} \tag{C-18}$$

C.3.1 Example 1 – Pressure Measurements

In this example, tire pressure is measured with a gauge. The procedure consists of taking a small sample of measurements and recording the average, standard deviation and sample size. This procedure is repeated five times. The resulting pressure data are listed below.

Sample Number	Average Pressure (lb$_f$/in^2)	Standard Deviation (lb$_f$/in^2)	Sample Size
1	31.7	0.6	3
2	32.3	0.8	5
3	32.0	1.0	3
4	30.5	1.3	4
5	32.7	0.6	3

The overall mean value is computed to be

$$\overline{P} = \frac{1}{18}\left(3 \times 31.7 + 5 \times 32.3 + 3 \times 32.0 + 4 \times 30.5 + 3 \times 32.7\right) \, \text{lb}_f / \text{in}^2$$

$$= \frac{572.7}{18} \, \text{lb}_f / \text{in}^2 = 31.8 \, \text{lb}_f / \text{in}^2$$

and the difference between the sampled mean values and the overall mean are

$$\overline{P}_1 - \overline{P} = 31.7 - 31.8 = -0.1 \, \text{lb}_f / \text{in}^2$$
$$\overline{P}_2 - \overline{P} = 32.3 - 31.8 = 0.5 \, \text{lb}_f / \text{in}^2$$
$$\overline{P}_3 - \overline{P} = 32.0 - 31.8 = 0.2 \, \text{lb}_f / \text{in}^2$$
$$\overline{P}_4 - \overline{P} = 30.5 - 31.8 = -1.3 \, \text{lb}_f / \text{in}^2$$
$$\overline{P}_5 - \overline{P} = 32.7 - 31.8 = 0.9 \, \text{lb}_f / \text{in}^2$$

The standard deviation of the sampled mean values relative to the overall mean is

$$s_b = \sqrt{\frac{1}{17}\left[3 \times (-0.1)^2 + 5 \times (0.5)^2 + 3 \times (0.2)^2 + 4 \times (-1.3)^2 + 3 \times (0.9)^2\right]} \, \text{lb}_f / \text{in}^2$$

$$= \sqrt{\frac{10.6}{17}} \, \text{lb}_f / \text{in}^2 = 0.8 \, \text{lb}_f / \text{in}^2.$$

The within sample sigma is computed to be

$$s_w = \sqrt{\frac{1}{17}\left[(3-1) \times 0.6^2 + (5-1) \times 0.8^2 + (3-1) \times 1.0^2 + (4-1) \times 1.3^2 + (3-1) \times 0.6^2\right]} \, \text{lb}_f / \text{in}^2$$

$$= \sqrt{\frac{11}{17}} \, \text{lb}_f / \text{in}^2 = 0.8 \, \text{lb}_f / \text{in}^2.$$

The standard deviation is then computed to be

$$s = \sqrt{(0.8)^2 + (0.8)^2} \, \text{lb}_f / \text{in}^2 = \sqrt{1.28} \, \text{lb}_f / \text{in}^2 = 1.13 \, \text{lb}_f / \text{in}^2$$

and the standard deviation in the mean of the sample mean values is computed to be

$$s_{\overline{P}} = \frac{1.13 \, \text{lb}_f / \text{in}^2}{\sqrt{5}} = 0.51 \, \text{lb}_f / \text{in}^2.$$

C.4   Outlier Testing

In the context of this document, an outlier is defined as a measured value that "appears" to be inconsistent with other values observed within a data sample. Statistically speaking, an outlier has a low probability of belonging to the same underlying distribution as other sampled values.

As such, however, an apparent outlier may just be an observed value located near the tail of the distribution.

Depending on the sample size, one or two outliers can significantly affect the calculated statistics by falsely increasing the standard deviation (i.e., distribution spread) or introducing a bias in the mean value. Consequently, the identification and possible exclusion of outliers from the calculation of sample statistics may be warranted.

There are various test methods available for identifying statistical outliers from data samples. Unfortunately, no single method or practice has gained universal acceptance. Similarly, no consensus exists regarding the exclusion of outliers from subsequent data analysis.[73]

The criteria for defining and identifying outliers can often be subjective. Therefore, the decision to exclude outliers from your sample statistics should be based on sufficient knowledge about the measurement process. It is also a good practice to report all sample data, including potential outliers and the method used to identify them.

C.4.1 Background

Most outlier tests are based on the evaluation of the relative deviation between the suspected outlier and the sample mean. There are several outlier tests based on the assumption that the sample data are normally distributed. These include Grubbs' test, Dixon's test, Rosner's test and Chauvenet's criterion.

If the data are not believed to follow a normal distribution, then non-parametric (i.e., distribution independent) tests can be applied. However, non-parametric outlier tests are not considered to be as reliable as parametric tests and often require sample sizes of 100 or more.

Grubb's test identifies one outlier at a time, thus requiring an iterative application. Dixon's test, Rosner's test and Chauvenet's criterion identify one or more outliers. Chauvenet's criterion has achieved relatively wide acceptance because it applies a simple, yet extremely effective non-parametric technique to identify potential outliers.

C.4.2 Chauvenet's Criterion[74]

Chauvenet's criterion defines acceptable scatter around a mean value $\bar{x}$ for a given sample of n readings and standard deviation s_n. It specifies that all points should be retained that fall within a band around the mean value that corresponds to a probability of $1 - 1 / 2n$.

The normal distribution is used to determine the number of sample standard deviations that relate to this probability. This "coverage factor" is obtained using the two-tailed inverse normal function $\Phi^{-1}()$

$$L_n = \Phi^{-1}\left(\frac{1 + P_n}{2}\right), \tag{C-19}$$

---

[73] In fact, the FDA guidance "Investigating Out of Specification (OOS) Test Results for Pharmaceutical Production" indicates that a chemical test result cannot be omitted with an outlier test, but a bioassy can be omitted. Content uniformity and dissolution testing are specific areas that prohibit outlier removal.

[74] Coleman, H. W. and Steele, W. G.: Experimentation and Uncertainty Analysis for Engineers, 2nd Edition, Wiley Interscience Publication, John Wiley & Sone, Inc., 1999, pp 34-37.

where

$$P_n = 1 - 1/2n.$$

Any points that lie outside $\bar{x} \pm L_n s_n$ are rejected.

C.5   Normality Testing

As previously discussed, the statistical analysis of samples is often based on the assumption that the data follow the normal distribution. Therefore, it is often necessary to assess whether the data are indeed normally distributed or a least approximately normally distributed. If the data are not normally distributed, then the following questions should be asked

- Is the apparent non-normality a result of potential outliers?
- Can the data be normalized via a transform function (e.g., log transform)?
- Should the data be evaluated using non-parametric (i.e., distribution-free) statistics?

C.5.1 Background

Both qualitative and quantitative methods can used to determine if the sampled data can be assumed to be normally distributed. Qualitative or graphical methods include the use of frequency histogram, normal probability and box-whisker plots. Quantitative or statistical methods include tests for skewness and kurtosis, the chi-squared test, the Kolmogorov-Smirnov test and the Shapiro-Wilk test, as well as variations of these tests.[75]

While graphical techniques provide a visual depiction of the data, their interpretation can be highly subjective, especially when the sample size is small (i.e., $n < 10$). Statistical tests provide more formal, objective methods for assessing whether the normal distribution provides an adequate description of the observed data.

Statistical normality tests typically include the following basic procedure:

1. A test statistic is calculated from the observed data.
2. Assuming the normal distribution is indeed applicable, the probability of obtaining the calculated test statistic is determined.
3. If the probability of obtaining the calculated test statistic is low (i.e., less than 0.05) then it is concluded that the normal distribution does not provide an adequate representation of the observed data. Conversely, if the probability is not low, then there is no evidence to reject the assumptions that the data are normally distributed.

    Note: The value set for the low probability is based on a user-defined confidence level (i.e., 90%, 95% or 99%). It is also important to understand that the outcome of a statistical test is highly dependent on the amount of data available. The larger

---

[75] For example, see Bain, L. J. and Engelhardt, M.: *Introduction to Probability and Mathematical Statistics*, Duxbury Press, 1992.

the sample size, the better the chances of rejecting (or accepting) the normal distribution assumption.

The chi-squared and Shapiro-Wilk tests provide the best means of determining whether or not the data are sampled from a normal distribution. The chi-squared method requires large data samples (i.e., $n \geq 50$). An advantage of the Shapiro-Wilk test is that it can be used for smaller sample sizes ($20 \leq n \leq 50$).

For samples of size 10 or more, statistical tests can also be performed to evaluate the skewness and kurtosis of the sample in comparison to what is expected of samples from a normally distributed population.

C.5.2 Skewness and Kurtosis Tests ($n \geq 10$)

Descriptive statistics, such as skewness and kurtosis, can provide relevant information about the normality of the data sample. Skewness is a measure of how symmetric the data distribution is about its mean. Kurtosis is a measure of the "peakedness" of the distribution.

If $x_1, x_2, \cdots, x_n$ are sampled values from a sample of size n with mean $\bar{x}$ and standard deviation s, the sample coefficient of skewness c_3 and coefficient of kurtosis c_4 are given by[76]

$$c_3 = \frac{\dfrac{1}{n-1}\sum_{i=1}^{n}(x_i - \bar{x})^3}{s^3} \tag{C-20}$$

and

$$c_4 = \frac{\dfrac{1}{n-1}\sum_{i=1}^{n}(x_i - \bar{x})^4}{s^4} \tag{C-21}$$

where

$$s = \sqrt{\frac{1}{n-1}\sum_{i=1}^{n}(x_i - \bar{x})^2} \, .$$

The coefficient of skewness for a normal distribution is 0 (i.e., there is no deviation from symmetry). The kurtosis of the normal distribution is 3. Consequently, if the skewness of the data sample differs significantly from 0, then it exhibits an asymmetric distribution. Similarly, if the kurtosis is significantly different from 3, then the distribution is either flatter or more peaked than the normal distribution.

C.5.3 Chi-square (χ^2) Test ($n \geq 50$)

The chi-squared goodness-of-fit test is based on the relative differences between observed frequencies from a histogram plot of the data and the theoretical frequencies predicted by the probability density function for the normal distribution.

---

[76] NIST/SEMATECH, e-Handbook of Statistical Methods, www.ITL.NIST.gov/div898/handbook /eda/section3/eda35b.htm.

182

C.5.4 Shapiro-Wilk Test ($20 \leq n \leq 50$)

The Shapiro-Wilk test for normality consists of computing a W statistic based on the tabulated coefficients, the sample standard deviation and sample size. A critical value W_α is also obtained from tabulated values for sample size n and significance level α, which is usually set equal to 0.10 or 0.05. The criteria for accepting or rejecting the normal distribution hypothesis is whether or not $W \geq W_\alpha$.

C.6 Sample Size Evaluation

As previously stated, sample size can affect results of normality and outlier tests. In fact, some test methods require a minimum sample size. More importantly, however, the size of a data sample can affect the computed sample mean $\bar{x}$, standard deviation s_x, and the standard deviation in the mean $s_{\bar{x}}$.

For example, consider the sample of AC voltage measurements given in Section C.1.1. If the measurement process stopped after the first 5 voltage measurements were collected, then

$$\bar{V} = \frac{1}{5}\left(115.5 + 116.0 + 116.5 + 114.3 + 115.3\right) V_{ac} = 115.5 \ V_{ac}$$

$$V_1 - \bar{V} = 115.5 - 115.5 = 0.0 \ V_{ac}$$
$$V_2 - \bar{V} = 116.0 - 115.5 = 0.5 \ V_{ac}$$
$$V_3 - \bar{V} = 116.5 - 115.5 = 1.0 \ V_{ac}$$
$$V_4 - \bar{V} = 114.3 - 115.5 = -1.2 \ V_{ac}$$
$$V_5 - \bar{V} = 115.3 - 115.5 = -0.2 \ V_{ac}$$

$$s_V = \sqrt{\frac{1}{4}\left[(0.0)^2 + (0.5)^2 + (1.0)^2 + (-1.2)^2 + (-0.2)^2\right]} V_{ac} = 0.83 \ V_{ac}$$

and

$$s_{\bar{V}} = \frac{0.83 \ V_{ac}}{\sqrt{5}} = 0.37 \ V_{ac}.$$

Comparison of the computed statistics obtained for the two sample sizes are shown below.

$n =$	5	15
$\bar{V} =$	115.5	115.9
$s_V =$	0.83	0.77
$s_{\bar{V}} =$	0.37	0.20

In general, the sample size should be sufficient to achieve the goal of data sampling, which is to make inferences about the population characteristics. Therefore, we must return to the question: "how large does the sample size need to be?"

A sample size evaluation method, based on the central limit theorem and probability theory, is discussed in Section C.6.1. This method provides a straightforward approach to the assessment of the minimum sample size required to achieve a specified maximum deviation between the sample mean and the population mean.

C.6.1 Methodology[77]

Let $x_1, x_2, \cdots, x_n$ represent repeat, independent unbiased measurements from a distribution with mean μ and standard deviation σ. According to the law of large numbers,[78] the sample average $\bar{x}$ for these measurements converges to μ in probability. Therefore, we can assume that is a good estimate of μ, if the sample size n is large.

The central limit theorem allows us to use the normal distribution to estimate the probability that the magnitude of the difference between $\bar{x}$ and μ is less than some maximum value c.

$$P(|\bar{x} - \mu| < c) = P(-c < \bar{x} - \mu < c) \tag{C-22}$$

To estimate P, we first note that the expectation value of $\bar{x}$ is μ, and the variance in $\bar{x}$ is σ^2/n. Then the variable

$$\frac{\bar{x} - \mu}{\sigma/\sqrt{n}}$$

is normally distributed with population mean = 0 and population variance = 1. Accordingly, we can write

$$P(-c < \bar{x} - \mu < c) = P\left(\frac{-c}{\sigma/\sqrt{n}} < \frac{\bar{x} - \mu}{\sigma/\sqrt{n}} < \frac{c}{\sigma/\sqrt{n}}\right) \tag{C-23}$$

$$= \Phi\left(\frac{c}{\sigma/\sqrt{n}}\right) - \Phi\left(\frac{-c}{\sigma/\sqrt{n}}\right)$$

$$= 2\Phi\left(\frac{c}{\sigma/\sqrt{n}}\right) - 1$$

where Φ is the normal distribution function. Equating this probability to a confidence level β for the condition $|\bar{x} - \mu| < c$, we have

$$2\Phi\left(\frac{c}{\sigma/\sqrt{n}}\right) - 1 = \beta \tag{C-24}$$

and

$$\frac{c}{\sigma/\sqrt{n}} = \Phi^{-1}\left(\frac{1+\beta}{2}\right) \tag{C-25}$$

---

[77] Rice, J.: *Mathematical Statistics and Data Analysis*, Duxbury Press, Belmont, 1995, page 172.

[78] The law of large numbers is a fundamental theorem of probability developed by Jacob Bernoulli circa 1713.

where Φ^{-1} is the inverse normal distribution function. Rearranging equation (C-25), we have

$$\sqrt{n} = (\sigma/c)\Phi^{-1}\left(\frac{1+\beta}{2}\right) \qquad \text{(C-26)}$$

In practice, we usually don't know the value of σ. Accordingly, we use the best available estimate. In many cases, this is the sample standard deviation s. With this substitution, we have

$$\sqrt{n} = (s/c)\Phi^{-1}\left(\frac{1+\beta}{2}\right) \qquad \text{(C-27)}$$

C.6.2 Example 1 – Evaluation using a Sample Standard Deviation

In this example, we will use the measurement sample listed below to estimate the minimum sample size needed to ensure that the sample mean will fall within 0.8 VAC of the population mean with 95% confidence level.

Measurement	AC Voltage
1	115.5
2	116.0
3	116.5
4	114.3
5	115.3
6	117.1

The sample mean is computed to be

$$\bar{V} = \frac{1}{6}(115.5 + 116.0 + 116.5 + 114.3 + 115.3 + 117.1)\,V_{ac}$$

$$= \frac{694.7\,V_{ac}}{6} = 115.8\,V_{ac}$$

and the differences between the measured values and the mean value are

$$V_1 - \bar{V} = 115.5 - 115.8 = -0.3\,V_{ac}$$
$$V_2 - \bar{V} = 116.0 - 115.8 = 0.2\,V_{ac}$$
$$V_3 - \bar{V} = 116.5 - 115.8 = 0.7\,V_{ac}$$
$$V_4 - \bar{V} = 114.3 - 115.8 = -1.5\,V_{ac}$$
$$V_5 - \bar{V} = 115.3 - 115.8 = -0.5\,V_{ac}$$
$$V_6 - \bar{V} = 117.1 - 115.8 = 1.3\,V_{ac}.$$

The standard deviation is

$$s_V = \sqrt{\frac{1}{5}\left[(-0.3)^2 + (0.2)^2 + (0.7)^2 + (-1.5)^2 + (-0.5)^2 + (1.3)^2\right]}\, V_{ac}$$

$$= \sqrt{\frac{4.81}{5}}\, V_{ac} = 0.98\, V_{ac}.$$

Applying equation (C-27), with $c = 0.8\ V_{ac}$ and $\beta = 0.95$, we have

$$\sqrt{n} = (0.98/0.8)\Phi^{-1}\left(\frac{1+0.95}{2}\right)$$

$$= 1.23 \times \Phi^{-1}(0.975)$$

$$= 1.23 \times 1.96$$

$$= 2.4$$

and $n = (2.4)^2 \cong 6$. Therefore, given our initial criteria, the existing sample size should be sufficient. However, if we had set $c = 0.6\ V_{ac}$ then

$$\sqrt{n} = (0.98/0.6)\Phi^{-1}\left(\frac{1+0.95}{2}\right)$$

$$= 1.63 \times \Phi^{-1}(0.975)$$

$$= 1.63 \times 1.96$$

$$= 3.2$$

and $n = (3.2)^2 \cong 10$. In this case, the existing sample size of 6 would not be sufficient.

C.6.3 Example 2 – Evaluation using a Population Standard Deviation

In this example, we will assume that a special temperature measurement test was conducted to collect a large data sample (i.e., 50 or more observations) to characterize the population standard deviation, σ. From analysis of the large data sample we obtained a value of $\sigma = 0.1$ °F.

We will use this estimation for the population standard deviation to economize the collection of future samples based on the following criteria

$$c = 0.05\ \text{°F} \quad \text{and} \quad \beta = 0.99.$$

Applying equation (C-26), we have

$$\sqrt{n} = (0.1/0.05)\Phi^{-1}\left(\frac{1+0.99}{2}\right)$$

$$= 2 \times \Phi^{-1}(0.995)$$

$$= 2 \times 2.576$$

$$= 5.15$$

and $n \cong (5.15)^2 \cong 27$.

However, if we lower our confidence level to $\beta = 0.95$, then

$$\sqrt{n} = (0.1/0.05)\Phi^{-1}\left(\frac{1+0.95}{2}\right)$$
$$= 2 \times \Phi^{-1}(0.975)$$
$$= 2 \times 1.96$$
$$= 3.92$$

and $n \cong (3.92)^2 \cong 15$.

Alternatively, we can use equation (C-24) to estimate the confidence level for the condition $|\bar{T} - \mu| < 0.05 \text{ }^\circ F$, given $\sigma = 0.1$ and $n = 10$.

$$\beta = 2\Phi\left(\frac{0.05}{0.1/\sqrt{10}}\right) - 1$$
$$= 2\Phi(1.58) - 1$$
$$= 2 \times 0.943 - 1$$
$$= 0.886.$$

In this case, there is a 88.6% probability that the value of $\bar{T}$ obtained from 10 repeat measurements would be within $\pm 0.05 \text{ }^\circ F$ of the population mean, μ.

APPENDIX D – ESTIMATING TYPE B DEGREES OF FREEDOM

The amount of information used to estimate the uncertainty in a given error source is called the *degrees of freedom*. The degrees of freedom is required, among other things, to employ an uncertainty estimate in computing confidence limits commensurate with some desired confidence level.

A Type A estimate is a standard deviation computed from a sample of data. In test or calibration, the sample standard deviation represents the uncertainty due to random error or repeatability accompanying a measurement. From the discussion in Appendix C, recall that the degrees of freedom for this uncertainty is given by

$$\nu = n - 1$$

where n is the sample size.

A Type B estimate is, by definition, an estimate obtained without recourse to a sample of data. Accordingly, for a Type B estimate, we don't have a sample size to work with. However, we can develop something analogous to a sample size by applying the method described herein.

This methodology was originally developed in 1997 by Dr. Howard Castrup to provide a rigorous approach for estimating Type B degrees of freedom. The method includes a formal structure for extracting information from the measurement experience of scientific or technical personnel. This information is used to calculate Type B uncertainty estimates and to approximate the degrees of freedom of the estimate.

D.1 Methodology

The approach used to estimate the degrees of freedom for Type B estimates begins with the relation proposed in the GUM.[79]

$$\nu \cong \frac{1}{2} \frac{u^2(x)}{\sigma^2[u(x)]} \tag{D-1}$$

The method for computing the variance[80] in the uncertainty, $\sigma^2[u(x)]$, is outlined in the following steps:

1. We generalize the equation for the Type B uncertainty estimate as

$$u_B = \frac{L}{\varphi(p)} \tag{D-2}$$

where L in the containment limit, p is the containment probability, and $\varphi(p)$ is defined as

---

[79] Equation G.3, Annex G of the GUM.

[80] In this document, the terms $\sigma^2[\,]$ and $\sigma^2()$ are equivalent to variance operator var().

188

$$\varphi(p) = \Phi^{-1}\left[(1+p)/2\right] \tag{D-3}$$

and the function $\Phi^{-1}[\cdot]$ is the inverse normal distribution function.

2. The error in the uncertainty, u_B, due to errors in L and p is estimated using a first order Taylor Series expansion.

$$
\begin{aligned}
\varepsilon_{u_B} &= \left(\frac{\partial u_B}{\partial L}\right)\varepsilon_L + \left(\frac{\partial u_B}{\partial p}\right)\varepsilon_p \\
&= \left(\frac{\partial u_B}{\partial L}\right)\varepsilon_L + \left(\frac{\partial u_B}{\partial \varphi}\right)\frac{d\varphi}{dp}\varepsilon_p \\
&= \frac{\varepsilon_L}{\varphi} - \frac{L}{\varphi^2}\frac{d\varphi}{dp}\varepsilon_p
\end{aligned}
\tag{D-4}
$$

where ε_L and ε_p are errors in L and p, respectively.

3. Assuming statistical independence between ε_L and ε_p, the variance in u_B is obtained using the variance addition rule.

$$\sigma^2(u_B) = \mathrm{var}(\varepsilon_{u_B}) = \frac{1}{\varphi^2}\mathrm{var}(\varepsilon_L) + \frac{L^2}{\varphi^4}\left(\frac{d\varphi}{dp}\right)^2 \mathrm{var}(\varepsilon_p) \tag{D-5}$$

By definition, the uncertainty of a quantity x is equal to the square root of the variance in the error in x.

$$u_x = \sqrt{\mathrm{var}(\varepsilon_x)}$$

Therefore, the variance in ε_L and ε_p can be expressed as

$$\mathrm{var}(\varepsilon_L) = u_L^2 \text{ and } \mathrm{var}(\varepsilon_p) = u_p^2.$$

Equation (D-5) can then be expressed as

$$\sigma^2(u_B) = \frac{u_L^2}{\varphi^2} + \frac{L^2}{\varphi^4}\left(\frac{d\varphi}{dp}\right)^2 u_p^2. \tag{D-6}$$

Dividing equation (D-6) by the square of equation (D-2), we get

$$\frac{\sigma^2(u_B)}{u_B^2} = \frac{u_L^2}{L^2} + \frac{1}{\varphi^2}\left(\frac{d\varphi}{dp}\right)^2 u_p^2 \tag{D-7}$$

The derivative in equation (D-7) is obtained from equation (D-3). We first establish that

$$\frac{1+p}{2} = \Phi[\varphi] = \frac{1}{\sqrt{2\pi}} \int_{-\infty}^{\varphi} e^{-\zeta^2/2} d\zeta \qquad \text{(D-8)}$$

where $\Phi[\cdot]$ is the probability density function for the normal distribution.

We next take the derivative of both sides of this equation with respect to p to get

$$\frac{1}{2} = \frac{1}{\sqrt{2\pi}} e^{-\varphi^2/2} \frac{d\varphi}{dp} \qquad \text{(D-9)}$$

and, finally,

$$\frac{d\varphi}{dp} = \sqrt{\frac{\pi}{2}} \, e^{\varphi^2/2}. \qquad \text{(D-10)}$$

Substituting equation (D-10) into equation (D-7) gives

$$\frac{\sigma^2(u_B)}{u_B^2} = \frac{u_L^2}{L^2} + \frac{1}{\varphi^2} \frac{\pi}{2} e^{\varphi^2} u_p^2$$

which, with the aid of equation (D-1), yields

$$v_B = \frac{1}{2} \left[\frac{u_L^2}{L^2} + \frac{1}{\varphi^2} \frac{\pi}{2} e^{\varphi^2} u_p^2 \right]^{-1}. \qquad \text{(D-11)}$$

D.2 Analysis Formats

In applying equation (D-11), we are confronted with the problem of obtaining u_L and u_p. These quantities can be estimated using any of the four formats described in the following subsections.

D.2.1 Format 1: % of Values

This format reads "Approximately $C\%$ ($\pm\Delta c\%$) of observed values have been found to lie within the limits $\pm L$ ($\pm\Delta L$)."

In this format, a technical expert is asked to provide the error limits $\pm\Delta L$ and $\pm\Delta c\%$. These limits are used to estimate u_A and u_p. The containment probability is

$$p = C / 100$$

where C is the percentage of values of y observed within $\pm L$.

If we assume that the errors in the estimates of L and p are approximately uniformly distributed within $\pm\Delta L$ and $\pm\Delta p = \pm\Delta c\% / 100$, respectively, then we can write

$$u_L^2 = \frac{(\Delta L)^2}{3} \tag{D-12}$$

and

$$u_p^2 = \frac{(\Delta p)^2}{3} \tag{D-13}$$

Use of the uniform distribution is appropriate here, since the ranges $\pm\Delta L$ and $\pm\Delta p$ can be considered analogous to "limits of resolution," for which the uniform distribution is applicable. This obviates the need for estimating confidence levels for ΔL and Δp. Any lack of rigor introduced by this tactic is felt as a third order effect and does not materially compromise the rigor of our final result. Note, however, that the minimum limits criterion, described in Appendix B, are still in effect.

Substituting equations (D-12) and (D-13) in equation (D-11) gives

$$\frac{\sigma^2(u_B)}{u_B^2} = \frac{(\Delta L)^2}{3L^2} + \frac{1}{\varphi^2}\frac{\pi}{2}e^{\varphi^2}\frac{(\Delta p)^2}{3} \ . \tag{D-14}$$

Using equation (D-14) in equation (D-1) yields an estimate for the degrees of freedom, v_B, for a Type B uncertainty estimate.

$$v_B \cong \frac{1}{2}\left(\frac{\sigma^2(u_B)}{u_B^2}\right)^{-1} \cong \frac{3\varphi^2 L^2}{2\varphi^2(\Delta L)^2 + \pi L^2 e^{\varphi^2}(\Delta p)^2} \tag{D-15}$$

If ΔL and Δp are set equal to zero, then the Type B degrees of freedom becomes infinite. Obviously, in most cases, it is not realistic to have infinite degrees of freedom for Type B uncertainty estimates. Therefore, it behooves us to attempt to apply whatever means we have at our disposal to obtain a sensible estimate for v_B.

D.2.2 Format 2: X out of N

This format reads "Approximately X out of N observed values have been found to lie within the limits $\pm L$ ($\pm\Delta L$)."

In this format, the containment probability is expressed as $p = X/N$, where N is the number of observations of a value and X is the number of values observed to fall within $\pm L$ ($\pm\Delta L$). The variance in L is obtained the same as in Format 1. The variance in the containment probability p can be obtained by taking advantage of the binomial character of p.

$$u_p^2 = \frac{p(1-p)}{N} \tag{D-16}$$

Substituting in equations (D-12) and (D-16) into equation (D-11) gives

191

$$\frac{\sigma^2(u_B)}{u_B^2} = \frac{(\Delta L)^2}{3L^2} + \frac{1}{\varphi^2}\frac{\pi}{2}e^{\varphi^2}\frac{p(1-p)}{N}.$$ (D-17)

Using equation (D-17) in equation (D-1) yields

$$v_B \cong \frac{1}{2}\frac{u_B^2}{\sigma^2(u_B)} \cong \frac{3\varphi^2 L^2}{2\varphi^2(\Delta L)^2 + 3\pi L^2 e^{\varphi^2}p(1-p)/N}.$$ (D-18)

D.2.3 Format 3: % of Cases

This format reads "Approximately C% of N observed values have been found to lie within the limits $\pm L$ ($\pm\Delta L$)."

This format is a variation of Format 2 in which the containment probability is stated in terms of a percentage C of the number of observations n, with $p = C/100$. The equation for estimating the degrees of freedom is the same as for Format 2:

$$v_B \cong \frac{3\varphi^2 L^2}{2\varphi^2(\Delta L)^2 + 3\pi L^2 e^{\varphi^2}p(1-p)/N}.$$ (D-19)

D.2.4 Format 4: % Range

This format reads "Between C_1% and C_2% of observed values have been found to lie between the limits $\pm L$ ($\pm\Delta L$)."

This format is a variation of Format 1 in which a range of values is given for the containment probability, $p = C/100$, where $C = (C_1 + C_2)$ and $\pm\Delta c = (C_2 - C_1)/2$. The equation for estimating the degrees of freedom is the same as for Format 1:

$$v_B \cong \frac{3\varphi^2 L^2}{2\varphi^2(\Delta L)^2 + \pi L^2 e^{\varphi^2}(\Delta p)^2}.$$ (D-20)

APPENDIX E – BAYESIAN ANALYSIS

Using Bayes theorem, methods were developed in the mid to late 1980s that enabled the analysis of false accept risk for unit-under-test (UUT) parameters, the estimation of both UUT parameter and measurement reference (MTE) biases, and the uncertainties in these biases. These methods have been referred to as Bayesian risk analysis methods or, simply, Bayesian analysis methods.

In applying Bayesian analysis methods, we can refine estimates of the UUT and MTE attribute biases and compute in-tolerance probabilities based on *a priori* knowledge and on measurement results obtained during testing or calibration.

The fundamentals of the Bayesian method are presented in the following sections. Derivations of the expressions used in this appendix are given in NASA *Measurement Quality Assurance Handbook* ANNEX 4 – *Estimation and Evaluation of Measurement Decision Risk*.

> **Note**: The Bayesian method described herein is applicable when parameter biases are normally distributed.

E.1 Bayes Theorem

In the 18th century, Reverend Thomas Bayes expressed the probability of any event, E_1, – given that a related event, E_2, has occurred – as a function of the probabilities of the two events occurring independently and the probability of both events occurring together.

$$P(E_1 \mid E_2) = \frac{P(E_1, E_2)}{P(E_2)} \tag{E-1}$$

where the joint probability $P(E_1, E_2)$ is defined as

$$P(E_1, E_2) = P(E_2 \mid E_1)P(E_1) = P(E_1 \mid E_2)P(E_2) \tag{E-2}$$

So, the conditional probability $P(E_1|E_2)$ can be expressed as

$$P(E_1 \mid E_2) = \frac{P(E_2 \mid E_1)P(E_1)}{P(E_2)} \tag{E-3}$$

Bayes' theorem proves to be of considerable value in computing measurement decision risks in test and calibration. Its derivation is simple and straightforward.

E.1.1 Joint Probability

In measurement decision risk analysis, we are often interested in the probability of two events occurring simultaneously. For example, we might want to know the probability that a UUT attribute is both in-tolerance and perceived as being in-tolerance. If we represent the event of an in-tolerance attribute as E_1 and the event of observing the attribute to be in-tolerance as E_2, then the joint probability for occurrence of E_1 and E_2 is written

$$P(E_1 \text{ and } E_2) = P(E_1, E_2). \tag{E-4}$$

E.1.1.1   Statistical Independence

If the occurrence of event E_1 and the occurrence of event E_2 bear no relationship to one another, they are called statistically independent. For example, E_1 may represent the outcome that an individual selected at random from within a group of males is 30 years old and E_2 may represent the event that his shoe size is 11.

It can be shown that, for statistically independent events,

$$P(E_1, E_2) = P(E_1)P(E_2) \tag{E-5}$$

Another important result derives from the probability that event E_1 will occur or event E_2 will occur. The appropriate relation is

$$P(E_1 \text{ or } E_2) = P(E_1) + P(E_2) - P(E_1, E_2) \tag{E-6}$$

Substituting equation (E-5) into equation (E-6) gives the relation for cases where E_1 and E_2 are independent.

$$P(E_1 \text{ or } E_2) = P(E_1) + P(E_2) - P(E_1)P(E_2) \tag{E-7}$$

E.1.1.2   Mutually Exclusive Events

On occasion, events are mutually exclusive. That is, they cannot occur together. A popular example is the tossing of a coin. Either heads will occur or tails will occur. They obviously cannot occur simultaneously. This means that $P(E_1, E_2) = 0$, and

$$P(E_1 \text{ or } E_2) = P(E_1) + P(E_2) \tag{E-8}$$

E.1.2   Conditional Probability

If the occurrence of E_2 is influenced by the occurrence of E_1, we say that E_1 and E_2 are conditionally related and that the probability of E_2 is conditional on event E_1. Conditional probabilities are written

$$P(E_2 \text{ given } E_1) = P(E_2 \mid E_1) \tag{E-9}$$

It can be shown that the joint probability for E_1 and E_2 can be expressed as

$$P(E_1, E_2) = P(E_1 \mid E_2)P(E_2) \tag{E-10}$$

Equivalently, we can also write

$$P(E_2, E_1) = P(E_2 \mid E_1)P(E_1) \tag{E-11}$$

Note that, since $P(E_1, E_2) = P(E_2, E_1)$, we have

$$P(E_1 \mid E_2)P(E_2) = P(E_2 \mid E_1)P(E_1) \qquad \text{(E-12)}$$

Rearranging equation (E-12), we have Bayes' theorem given in equation (E-3).

$$P(E_1 \mid E_2) = \frac{P(E_2 \mid E_1)P(E_1)}{P(E_2)}$$

E.2   Risk Analysis for a Measured Variable

The procedure for applying Bayesian analysis methods to perform risk analysis for a measured attribute or parameter is as follows:

1. Assemble all relevant *a priori* knowledge, such as the tolerance limits for the UUT attribute, the tolerance limits for the MTE attribute, the in-tolerance probabilities for each attribute and the uncertainty of the measurement process.

2. Perform a measurement or set of measurements. This may consist either of measuring the UUT attribute with the MTE attribute, measuring the MTE attribute with the UUT attribute or using both attributes to measure a common artifact.

3. Estimate the UUT attribute and MTE attribute biases using Bayesian analysis methods.

4. Compute uncertainties in the bias estimates.

5. Act on the results. Report the biases and bias uncertainties, along with in-tolerance probabilities for the attributes, or adjust each attribute to correct the estimated biases, as appropriate.

E.3   *A priori* Knowledge

The *a priori* knowledge for a Bayesian analysis may include several kinds of information. For example, if the UUT attribute is the pressure of an automobile tire, such knowledge may include a rigorous projection of the degradation of the tire's pressure as a function of time since the tire was last inflated or a SWAG estimate based on the appearance of the tire's lateral bulge. However *a priori* knowledge is obtained, it should lead to the following quantities:

- Estimates of the uncertainties in the biases of both the UUT attribute and MTE attribute. These estimates may be obtained heuristically from containment limits and containment probabilities or by other means, if applicable.

- An estimate of the uncertainty in the measurement process, accounting for all error sources.

E.4   Post-Test Knowledge

The post-test knowledge in a Bayesian analysis consists of the results of measurement. As stated earlier, these results may be in the form of a measurement or a set of measurements. The measurements may be the result of readings provided by the MTE attribute from measurements of the UUT attribute, readings provided by the UUT attribute from measurements of the MTE attribute, or readings provided by both the UUT attribute and MTE attribute, taken on a common artifact.

E.5   Bias Estimates

Initial UUT attribute and MTE attribute biases are estimated using the method described in Chapter 8. The method encompasses cases where a measurement sample is taken by either the UUT attribute, the MTE attribute or both. The variables are given in Table E-1.

Table E-1. Bayesian Estimation Variables

Variable	Description
$e_{UUT,b}$	the UUT attribute bias at the time of calibration
$u_{UUT,b}$	the UUT attribute bias standard uncertainty
δ	a measurement (estimate) of $e_{UUT,b}$ obtained through calibration.
$e_{MTE,b}$	the MTE attribute bias at the time of calibration
$u_{MTE,b}$	the MTE attribute bias standard uncertainty
u_{cal}	the uncertainty in the UUT attribute calibration process, as defined in Chapter 8.
$-L_1$ and L_2	the lower and upper UUT attribute tolerance limits
$-l_1$ and l_2	the lower and upper MTE attribute tolerance limits

E.5.1   Refinement of the UUT Bias Estimate

Employing the nomenclature listed in Table E-2 and the Bayes' relation given in equation (E-3), the conditional distribution of $e_{UUT,b}$ given a value of δ is defined as

$$f(e_{UUT,b} \mid \delta) = \frac{f(\delta \mid e_{UUT,b}) f(e_{UUT,b})}{f(\delta)} \tag{E-13}$$

where the probability density function for δ is

$$\begin{aligned} f(\delta) &= \int_{-\infty}^{\infty} f(\delta, e_{UUT,b}) de_{UUT,b} \\ &= \int_{-\infty}^{\infty} f(\delta \mid e_{UUT,b}) f(e_{UUT,b}) de_{UUT,b}. \end{aligned} \tag{E-14}$$

For normally distributed values of $e_{UUT,b}$ and δ,

$$f(e_{UUT,b}) = \frac{1}{\sqrt{2\pi} u_{UUT,b}} e^{-e_{UUT,b}^2 / 2u_{UUT,b}^2} \tag{E-15}$$

and

$$f(\delta \mid e_{UUT,b}) = \frac{1}{\sqrt{2\pi} u_{cal}} e^{-(\delta - e_{UUT,b})^2 / 2u_{cal}^2} \tag{E-16}$$

Table E-2. Risk Analysis Probability Density Functions

pdf	Description
$f(e_{UUT,b})$	pdf for the UUT bias at the time of calibration
$f(\delta)$	pdf for the measurement result
$f(\delta, e_{UUT,b})$	pdf for the joint distribution of δ and $e_{UUT,b}$
$f(\delta \mid e_{UUT,b})$	pdf for the conditional distribution of δ given a value of $e_{UUT,b}$
$f(e_{UUT,b} \mid \delta)$	pdf for the conditional distribution of $e_{UUT,b}$ given a value of δ
$f(e_{MTE,b} \mid \delta)$	pdf for the conditional distribution of $e_{MTE,b}$ given a value of δ

Note: The pdf designations in Table E-2 are consistent with those used in *NASA Measurement Quality Assurance Handbook ANNEX 4 – Estimation and Evaluation of Measurement Decision Risk.*

Substituting equations (E-15) and (E-16) into equation (E-14) gives

$$
\begin{aligned}
f(\delta) &= \frac{1}{2\pi u_{cal} u_{UUT,b}} \int_{-\infty}^{\infty} e^{-(\delta - e_{UUT,b})^2 / 2u_{cal}^2} \times e^{-e_{UUT,b}^2 / 2u_{UUT,b}^2}\, de_{UUT,b} \\
&= \frac{1}{2\pi u_{cal} u_{UUT,b}} \int_{-\infty}^{\infty} e^{\left[-(\delta - e_{UUT,b})^2 / 2u_{cal}^2\right] + \left[-e_{UUT,b}^2 / 2u_{UUT,b}^2\right]}\, de_{UUT,b}
\end{aligned}
\tag{E-17}
$$

Evaluation of the exponential argument is provided in equation (E-18).

$$
\begin{aligned}
\text{arg} &= \left[-(\delta - e_{UUT,b})^2 / 2u_{cal}^2\right] + \left[-e_{UUT,b}^2 / 2u_{UUT,b}^2\right] \\
&= -\frac{1}{2}\left[\frac{-(\delta - e_{UUT,b})^2}{u_{cal}^2} + \frac{e_{UUT,b}^2}{u_{UUT,b}^2}\right] \\
&= -\frac{1}{2}\left[\left(\frac{1}{u_{cal}^2} + \frac{1}{u_{UUT,b}^2}\right) e_{UUT,b}^2 - 2\frac{\delta}{u_{cal}^2} e_{UUT,b} + \frac{\delta^2}{u_{cal}^2}\right] \\
&= -\frac{1}{2}\left[\frac{u_{cal}^2 + u_{UUT,b}^2}{u_{cal}^2 u_{UUT,b}^2} e_{UUT,b}^2 - 2\frac{\delta}{u_{cal}^2} e_{UUT,b} + \frac{\delta^2}{u_{cal}^2}\right]
\end{aligned}
\tag{E-18}
$$

Defining the combined uncertainty

$$
u_A = \sqrt{u_{cal}^2 + u_{UUT,b}^2}
\tag{E-19}
$$

equation (E-18) becomes

$$\text{arg} = -\frac{1}{2}\left[\frac{u_A^2}{u_{cal}^2 u_{UUT,b}^2}\left(e_{UUT,b}^2 - 2\frac{u_{UUT,b}^2}{u_A^2}\delta e_{UUT,b}\right) + \frac{\delta^2}{u_{cal}^2}\right]$$

$$= -\frac{1}{2}\left[\frac{u_A^2}{u_{cal}^2 u_{UUT,b}^2}\left(e_{UUT,b} - \frac{u_{UUT,b}^2}{u_A^2}\delta\right)^2 + \frac{\delta^2}{u_{cal}^2} - \frac{u_{UUT,b}^2}{u_{cal}^2 u_A^2}\delta^2\right]$$

$$= -\frac{1}{2}\left[\frac{u_A^2}{u_{cal}^2 u_{UUT,b}^2}\left(e_{UUT,b} - \frac{u_{UUT,b}^2}{u_A^2}\delta\right)^2 + \frac{\delta^2}{u_{cal}^2}\left(\frac{u_A^2 - u_{UUT,b}^2}{u_A^2}\right)\right]$$

$$= -\frac{1}{2}\left[\frac{u_A^2}{u_{cal}^2 u_{UUT,b}^2}\left(e_{UUT,b} - \frac{u_{UUT,b}^2}{u_A^2}\delta\right)^2 + \frac{\delta^2}{u_A^2}\right]$$

(E-20)

Substituting the variables

$$\gamma = \frac{u_{cal}u_{UUT,b}}{u_A} \quad \text{and} \quad \kappa = \frac{u_{UUT,b}^2}{u_A^2}$$

into equation (E-20), the exponential argument can be written as

$$\text{arg} = -\frac{1}{2}\left[\frac{\left(e_{UUT,b} - \kappa\delta\right)^2}{\gamma^2} + \frac{\delta^2}{u_A^2}\right]$$

(E-21)

Finally, defining the variable

$$\zeta = \frac{e_{UUT,b} - \kappa\delta}{\gamma}$$

where $de_{UUT,b} = \gamma d\zeta$, equation (E-17) can be written

$$f(\delta) = \frac{e^{-\delta^2/2u_A^2}}{2\pi u_{cal}u_{UUT,b}} \int_{-\infty}^{\infty} e^{-(e_{UUT,b}-k\delta)^2/2\gamma^2}\, de_{UUT,b}$$

$$= \frac{\gamma e^{-\delta^2/2u_A^2}}{2\pi u_{cal}u_{UUT,b}} \int_{-\infty}^{\infty} e^{-\zeta^2/2}\, d\zeta$$

$$= \frac{\gamma e^{-\delta^2/2u_A^2}}{2\pi u_{cal}u_{UUT,b}}\sqrt{2\pi} = \frac{\gamma e^{-\delta^2/2u_A^2}}{\sqrt{2\pi}u_{cal}u_{UUT,b}} = \frac{\dfrac{u_{cal}u_{UUT,b}}{u_A} e^{-\delta^2/2u_A^2}}{\sqrt{2\pi}u_{cal}u_{UUT,b}}$$ (E-22)

$$= \frac{1}{\sqrt{2\pi}u_A} e^{-\delta^2/2u_A^2}$$

Substituting equations (E-15), (E-16) and (E-22) into equation (E-13), yields

$$f(e_{UUT,b}\mid\delta) = \frac{u_A}{\sqrt{2\pi}u_{cal}u_{UUT,b}} e^{-\left[(\delta-e_{UUT,b})^2/2u_{cal}^2 + e_{UUT,b}^2/2u_{UUT,b}^2 - \delta^2/2u_A^2\right]}$$

$$= \frac{1}{\sqrt{2\pi}u_\beta} e^{-(e_{UUT,b}-\beta)^2/2u_\beta^2},$$ (E-23)

where

$$\beta = \kappa\delta = \frac{u_{UUT,b}^2}{u_A^2}\delta$$ (E-24)

and

$$u_\beta = \frac{u_{UUT,b}u_{cal}}{u_A}$$ (E-25)

Given these results, along with the properties of the normal distribution, we see that β is the refined estimate for $e_{UUT,b}$ and u_β is the estimated bias uncertainty.

$$\text{UUT Attribute Bias} = \beta = \frac{u_{UUT,b}^2}{u_A^2}\delta$$ (E-26)

and

$$\text{UUT Attribute Bias Uncertainty} = u_\beta = \frac{u_{UUT,b}u_{cal}}{u_A}$$ (E-27)

E.5.2   Refinement of the MTE Bias Estimate

With the Bayesian method, calibration results can be used to obtain an estimate of the bias of the calibration MTE attribute and the uncertainty in this estimate. This is accomplished by imagining that the UUT is calibrating the MTE. We begin by replacing $u_{UUT,b}$ with $u_{MTE,b}$ and δ with $-\delta$ in equation (E-26).

$$\text{MTE Attribute Bias} = \alpha = -\frac{u_{MTE,b}^2}{u_A^2}\delta \qquad \text{(E-28)}$$

The first step in estimating the MTE attribute bias uncertainty is to define a new uncertainty term.

$$u_{process} = \sqrt{u_{cal}^2 - u_{MTE,b}^2} \qquad \text{(E-29)}$$

Next, a calibration uncertainty is defined that would apply if the UUT were calibrating the MTE.

$$u'_{cal} = \sqrt{u_{UUT,b}^2 + u_{process}^2} \qquad \text{(E-30)}$$

Using this quantity in equation (E-27) yields the MTE attribute bias uncertainty.

$$\text{MTE Attribute Bias Uncertainty} = u_\alpha = \frac{u_{MTE,b}u'_{cal}}{u_A} \qquad \text{(E-31)}$$

E.6   In-Tolerance Probabilities

The in-tolerance probabilities for the UUT and MTE attributes are estimated by integrating the appropriate pdf over the corresponding upper and lower tolerance limits.

E.6.1   UUT Attribute In-Tolerance Probability

An estimate of the UUT attribute in-tolerance probability $P_{UUT,in}$ is obtained by integrating $f(e_{UUT,b}\,|\,\delta)$ from $-L_1$ to L_2.

$$
\begin{aligned}
P_{UUT,in} &= \frac{1}{\sqrt{2\pi}u_\beta}\int_{-L_1}^{L_2} e^{-(e_{UUT,b}-\beta)^2/2u_\beta}\,de_{UUT,b} \\[2mm]
&= \Phi\!\left(\frac{L_1+\beta}{u_\beta}\right) + \Phi\!\left(\frac{L_2-\beta}{u_\beta}\right) - 1
\end{aligned}
\qquad \text{(E-32)}
$$

where β is given in equation (E-26).

E.6.2   MTE Attribute In-Tolerance Probability

Since we have the necessary expressions at hand, we can also estimate the in-tolerance probability of the MTE attribute, $P_{MTE,in}$. This probability is obtained by integrating the pdf $f(e_{MTE,b}\,|\,\delta)$ from $-l_1$ to l_2.

$$P_{MTE,in} = \frac{1}{\sqrt{2\pi}u_\alpha} \int\limits_{-L_1}^{L_2} e^{-(e_{MTE,b}-\alpha)^2/2u_\alpha^2} de_{MTE,b}$$

$$= \Phi\left(\frac{l_1+\alpha}{u_\alpha}\right) + \Phi\left(\frac{l_2-\alpha}{u_\alpha}\right) - 1$$

(E-33)

where α is given in equation (E-28).

APPENDIX F – FORCE GAUGE ANALYSIS EXAMPLE

The purpose of this analysis is to estimate and report the uncertainty in the calibration result obtained for an applied forced. The force measurement uncertainty is estimated using the analysis procedure discussed in Chapter 5.

F.1   Measurement Process Overview

A Chatillon model DGGS-250G digital force gauge is calibrated using a weight set manufactured by Rice Lake Bearing Inc. The force gauge has a full scale output of 250 g-force or 8 oz-force (ozf) and a specified accuracy of $\pm 0.15\%$ FS ± 1 LSC (least significant character or count).[81] The digital display resolution of the force gauge is specified as 0.005 oz-force.

When specifications are reported as $\pm L_1 \pm L_2$, they are typically combined in root sum square to obtain the total specification limits, $\pm L$.

$$\pm L = \pm\sqrt{L_1^2 + L_2^2}$$

However, when contacted for verification, Chatillon technical support personnel stated that the accuracy specifications for their DGGS-250G digital force gage should be added. Therefore, the total specification limits are computed to be

$$\pm L = \pm\left(L_1 + L_2\right)$$
$$= \pm\left(8\,\text{ozf} \times \frac{0.15}{100} + 0.005\,\text{ozf}\right)$$
$$= \pm\left(0.012 + 0.005\right)\text{ozf}$$
$$= \pm 0.017\,\text{ozf}.$$

> **Note**: Given the inconsistencies in which equipment specifications are reported, it is always a good practice to seek additional manufacturer clarification to ensure their proper interpretation and application.

The force gauge is mounted on a calibration base and oriented in either the tension or compression mode. The total applied force is obtained by attaching a combination of calibration weights. In the tension mode, the calibration weights are hung from the measurement shaft of the force gauge. In the compression mode, the calibration weights are placed on the measurement shaft

The assigned mass of each calibration weight is corrected for local gravity and air buoyancy to determine the apparent force.[82]

$$F = m \times cf \qquad\qquad (\text{F-1})$$

where

---

[81] Chatillon DFGS Series Specification data sheet, SS-FM-3112-1101, November 2001, Ametek, Inc.

[82] P. King: "Determining Force from Assigned Mass Values," Wyle Interoffice Correspondence 5321-05-080, July 6, 2005.

$$F = \text{applied force, oz-force}$$
$$m = \text{applied mass, oz}$$
$$cf = \text{conversion factor, oz-force/oz}$$

The primary purpose of the calibration is to obtain an estimate of the bias of the unit under test (UUT) force gage. The calibration result is the difference between the average force gauge reading and the applied force. This difference is denoted by the variable δ and defined by

$$\delta = \bar{x} - F \qquad\qquad (F\text{-}2)$$

where $\bar{x}$ is the average force gauge reading.

Calibration data[83] for the tension and compression modes are listed in Tables F-1 and F-2, respectively.

Table F-1. Tension Mode Calibration Data

Applied Mass (oz)	Applied Force (ozf)	Force Gauge Reading Run1 (ozf)	Force Gauge Reading Run2 (ozf)	Force Gauge Reading Average (ozf)	Measured Difference δ (ozf)	Force Gauge Specification Limits (ozf)
0.0000	0.0000	0.000	0.000	0.000	0.0000	± 0.017
1.6000	1.5974	1.595	1.600	1.598	0.0001	± 0.017
3.2000	3.1948	3.195	3.195	3.195	0.0002	± 0.017
4.8000	4.7922	4.790	4.795	4.793	0.0003	± 0.017
6.4000	6.3897	6.385	6.380	6.388	-0.0022	± 0.017
8.0000	7.9871	7.985	7.985	7.985	-0.0021	± 0.017

Table F-2. Compression Mode Calibration Data

Applied Mass (oz)	Applied Force (ozf)	Force Gauge Reading Run1 (ozf)	Force Gauge Reading Run2 (ozf)	Force Gauge Reading Average (ozf)	Measured Difference δ (ozf)	Force Gauge Specification Limits (ozf)
0.0000	0.0000	0.000	0.000	0.000	0.0000	± 0.017
1.6000	1.5974	1.595	1.595	1.595	-0.0024	± 0.017
3.2000	3.1948	3.195	3.195	3.195	0.0002	± 0.017
4.8000	4.7922	4.790	4.790	4.790	-0.0022	± 0.017
6.4000	6.3897	6.385	6.385	6.385	-0.0047	± 0.017
8.0000	7.9871	7.985	7.985	7.985	-0.0021	± 0.017

If the value of δ falls outside of the specified tolerance limits,[84] then the UUT bias is typically deemed to be out-of-tolerance (OOT) or noncompliant.

However, errors in the calibration process can result in an incorrect OOT assessment (false-reject) or incorrect in-tolerance assessment (false-accept). The relationship between the calibration result, δ, and the true UUT bias, $e_{UUT,b}$, is generally expressed as

---

[83] Wyle Laboratories Calibration Data Sheet Number M64118 11Aug08

[84] Since the tolerance limits constitute the maximum permissible difference or deviation, they should be expressed in units that are consistent with those measured during calibration.

$$\delta = e_{UUT,b} + \varepsilon_{cal} . \tag{F-3}$$

The probability that the UUT bias is in-tolerance is based on the calibration result and its associated uncertainty. Therefore, all relevant calibration error sources must be identified and combined in a way that yields viable uncertainty estimates.

F.2 Uncertainty Analysis Procedure

The purpose of this analysis is to estimate and report the total uncertainty in δ for each force applied during the calibration. The uncertainty in δ is determined by applying the variance operator to equation (F-3) and taking the square root.

$$
\begin{aligned}
u_\delta &= \sqrt{\text{var}\left(\delta\right)} = \sqrt{\text{var}\left(e_{UUT,b} + \varepsilon_{cal}\right)} \\
&= \sqrt{\text{var}\left(\varepsilon_{cal}\right)}
\end{aligned}
\tag{F-4}
$$

The force gauge calibration error, ε_{cal}, is defined as

$$\varepsilon_{cal} = \varepsilon_F + \varepsilon_{res} + \varepsilon_{rep} \tag{F-5}$$

where

ε_F = error in the applied force
ε_{res} = force gauge resolution error
ε_{rep} = repeatability or random error

Note: Operator bias is not considered relevant to this analysis.

As shown in equation (F-1), the applied force is a function of the applied mass and the conversion factor. Consequently, the error in the applied force is defined as

$$\varepsilon_F = c_m \varepsilon_m + c_{cf} \varepsilon_{cf} \tag{F-6}$$

where

ε_m = error in the value of the calibration weight(s)
ε_{cf} = error in the gravitational and air buoyancy correction factor
c_m, c_{cf} = sensitivity coefficients

Substituting equation (F-6) into (F-5), the calibration error equation can be expressed as

$$\varepsilon_{cal} = c_m \varepsilon_m + c_{cf} \varepsilon_{cf} + \varepsilon_{res} + \varepsilon_{rep} . \tag{F-7}$$

Brief descriptions of the applicable calibration error sources are provided in the following subsections.

F.2.1 Calibration Weight (ε_m)

A Chatillon model 0.1OZ-1LB-F weight set is used to calibrate the force gauge. The mass values and tolerance limits of the weight set are listed in Table F-3. The applied mass is obtained by using a combination of weights, $m = m_1 + m_2 + m_3$. Therefore, the error in the total applied mass is expressed as

$$\varepsilon_m = \varepsilon_{m_1} + \varepsilon_{m_2} + \varepsilon_{m_3} \qquad\qquad (F\text{-}8)$$

In this analysis, the errors contained within the mass tolerance limits are assumed to follow a normal distribution. The tolerance limits are also assumed to represent 95%confidence limits.

Table F-3. Calibration Data for Weight Set[85]

Weight ID	Nominal Weight (oz)	Measured Weight (oz)	Deviation from Nominal (oz)	Tolerance Limits[86] (oz)
	0.1	0.09999	-0.00001	± 0.00005
	0.2	0.20003	0.00003	± 0.00006
0.3	0.3	0.30000	0.00000	± 0.00006
½	0.5	0.50003	0.00003	± 0.00010
1	1	1.00011	0.00011	± 0.00019
2	2	2.00014	0.00014	± 0.00039
4	4	4.00035	0.00035	± 0.00081
8	8	8.00052	0.00052	± 0.00159
16	16	16.00052	0.00052	± 0.00247

F.2.2 Conversion Factor (ε_{cf})

The factor for converting the applied mass to force is 0.9983830 oz-force/oz. The correction factor accounts for local gravity and air buoyancy. The expanded uncertainty for the correction factor is estimated to be ± 6 ppm or ± 5.99 × 10^{-6} oz-force/oz. These tolerance limits represent a coverage factor of $k = 2$ and the associated error distribution is characterized by the normal distribution.

F.2.3 Digital Resolution (ε_{res})

As previously stated, the digital display resolution of the Chatillon force gauge is specified as 0.005 oz-force. Therefore, the resolution error limits are ± 0.0025 oz-force (i.e., ± half the resolution). These limits represent 100% containment limits for a uniformly distributed error source.

F.2.4 Repeatability (ε_{rep})

The calibration process for electronic force gauges like the Chatillon model DGGS unit does not include steps for obtaining repeat measurements. A special test was conducted on a similar force gage to assess the repeatability associated with the calibration equipment, laboratory environmental conditions and other procedural steps. Ten repeat measurements were made at four different applied force values. The resulting data are listed in Table F-4.

The data indicate that any variation in the force gauge readings is less than the display resolution. Therefore, repeatability is not included as a source of uncertainty for this analysis.

---

[85] Wyle Calibration Data Sheet M59578_11Jun08.
[86] NIST Handbook 105-1, *Specifications and Tolerances for Reference Standards and Field Standard Weights and Measures, 1. Specifications and Tolerances for Field Standard Weights (NIST Class F).*

Table F-4. Repeatability Data for DFGS-2 Force Gage[87]

Nominal Mass (lb)	0.0000	0.4000	1.2000	2.0000
Nominal Force (lbf)	0.0000	0.3994	1.1981	1.9968
Run 1 UUT Reading	0.000	0.399	1.198	1.995
Run 2 UUT Reading	0.000	0.399	1.197	1.995
Run 3 UUT Reading	0.000	0.399	1.197	1.995
Run 4 UUT Reading	0.000	0.399	1.197	1.995
Run 5 UUT Reading	0.000	0.399	1.197	1.995
Run 6 UUT Reading	0.000	0.399	1.197	1.995
Run 7 UUT Reading	0.000	0.399	1.197	1.995
Run 8 UUT Reading	0.000	0.399	1.197	1.995
Run 9 UUT Reading	0.000	0.399	1.197	1.995
Run 10 UUT Reading	0.000	0.399	1.197	1.995
Average UUT Reading	0.000	0.399	1.197	1.995
Standard Deviation of UUT Reading	0.000	0.000	0.000	0.000

The revised error model for the force gauge calibration is given in equation (F-9).

$$\varepsilon_{cal} = c_m \varepsilon_m + c_{cf} \varepsilon_{cf} + \varepsilon_{res} \tag{F-9}$$

Applying the variance operator to equation (F-9) gives

$$
\begin{aligned}
\text{var}\left(\varepsilon_{cal}\right) &= \text{var}\left(c_m \varepsilon_m + c_{cf} \varepsilon_{cf} + \varepsilon_{res}\right) \\
&= c_m^2 \, \text{var}\left(\varepsilon_m\right) + c_{cf}^2 \, \text{var}\left(\varepsilon_{cf}\right) + \text{var}\left(\varepsilon_{res}\right) + c_m c_{cf} \, \text{cov}\left(\varepsilon_m, \varepsilon_{cf}\right) \\
&\quad + c_m \, \text{cov}\left(\varepsilon_m, \varepsilon_{res}\right) + c_{cf} \, \text{cov}\left(\varepsilon_{cf}, \varepsilon_{res}\right)
\end{aligned}
\tag{F-10}
$$

where cov() terms account for the covariance between pairs of error sources. Covariance is a statistical assessment of the mutual dependence of the errors. The covariances can have inconvenient physical dimensions, so the correlation coefficient is often used instead. For example, the correlation coefficient for ε_m and ε_{cf} is defined as

$$\rho_{\varepsilon_m \varepsilon_{cf}} = \frac{\text{cov}(\varepsilon_m, \varepsilon_{cf})}{u_{\varepsilon_m} u_{\varepsilon_{cf}}} \tag{F-11}$$

where u_{ε_m} and $u_{\varepsilon_{cf}}$ are the uncertainties in ε_m and ε_{cf}, respectively. Therefore, equation (F-10) can be expressed as

---

[87] Repeatability data for Chatillon Model DFGS-2 Digital Force Gage.

$$\text{var}\left(\varepsilon_{cal}\right) = c_m^2 \, \text{var}\left(\varepsilon_m\right) + c_{cf}^2 \, \text{var}\left(\varepsilon_{cf}\right) + \text{var}\left(\varepsilon_{res}\right) + c_m c_{cf} \rho_{\varepsilon_m \varepsilon_{cf}} u_{\varepsilon_m} u_{\varepsilon_{cf}}$$
$$+ c_m \rho_{\varepsilon_m \varepsilon_{res}} u_{\varepsilon_m} u_{\varepsilon_{res}} + c_{cf} \rho_{\varepsilon_{cf} \varepsilon_{res}} u_{\varepsilon_{cf}} u_{\varepsilon_{res}} \tag{F-12}$$

There are no correlations between error sources, so equation (F-12) can be simplified.

$$\text{var}\left(\varepsilon_{cal}\right) = c_m^2 \, \text{var}\left(\varepsilon_m\right) + c_{cf}^2 \, \text{var}\left(\varepsilon_{cf}\right) + \text{var}\left(\varepsilon_{res}\right) \tag{F-13}$$

The variance terms in equation (F-13) are equivalent to the square of the uncertainty in the corresponding error (e.g., $\text{var}(\varepsilon_m) = u_{\varepsilon_m}^2$). So, the uncertainty equation for δ can be rewritten in terms of the individual measurement process uncertainties and their associated sensitivity coefficients.

$$u_\delta = \sqrt{c_m^2 u_{\varepsilon_m}^2 + c_{cf}^2 u_{\varepsilon_{cf}}^2 + u_{\varepsilon_{res}}^2} \tag{F-14}$$

The partial derivative equations used to compute the sensitivity coefficients are listed below.

$$c_m = \frac{\partial F}{\partial m} = cf \qquad c_{cf} = \frac{\partial F}{\partial cf} = m$$

The measurement process uncertainties are estimated from the specification limits, containment probability (i.e., confidence level) and the inverse error distribution function.

The force gauge digital resolution uncertainty is estimated using $\pm\,0.0025$ oz-force error limits, the inverse uniform distribution function and a 1.00 containment probability (100% confidence level).

$$u_{res} = \frac{0.0025 \text{ oz-force}}{\sqrt{3}} = \frac{0.0025 \text{ oz-force}}{1.732} = 0.00144 \text{ oz-force}.$$

The uncertainty in the conversion coefficient is estimated using the expanded uncertainty limits, the inverse normal distribution function, Φ^{-1}, and a 0.9545 containment probability (95.45% confidence level).

$$u_{cf} = \frac{5.99 \times 10^{-6} \, \frac{\text{ozf}}{\text{oz}}}{\Phi^{-1}\left(\frac{1+0.9545}{2}\right)} = \frac{5.99 \times 10^{-6} \, \frac{\text{ozf}}{\text{oz}}}{2.000} = 2.995 \times 10^{-6} \, \frac{\text{ozf}}{\text{oz}}$$

Note: The digital resolution and conversion factor uncertainties do not change over the range of applied masses.

The uncertainty in the applied mass is based on the combined weight standards used and their associated tolerance limits. The uncertainty for an applied mass of 1.6 oz is computed for

207

illustration purposes. Three weight set masses are used: 1 oz, 0.5 oz and 0.1 oz. From Table F-3, the total tolerance limits for the 1.6 oz applied mass are

$$\pm L_m = \pm (0.00005 + 0.00010 + 0.00019)\ oz = \pm 0.00034\ oz.$$

The uncertainty in the applied mass is estimated using the inverse normal distribution function and a 0.95 containment probability (95% confidence level). The uncertainty due to the error in the 1.6 oz applied mass is estimated to be

$$u_m = \frac{0.00034\ \text{oz-force}}{\Phi^{-1}\left(\frac{1+0.95}{2}\right)} = \frac{0.00034\ \text{oz-force}}{1.9600} = 0.000173\ \text{oz-force}$$

The estimated uncertainties and sensitivity coefficients for each parameter are summarized in Table F-5. The component uncertainties are the product of the standard uncertainty and the sensitivity coefficient.

Table F-5. Estimated Uncertainties for Force Gauge Calibration at Applied Mass = 1.6 oz

Error Source	± Error Limits	Conf. Level	Standard Uncertainty	Sensitivity Coefficient	Component Uncertainty
ε_m	± 0.00034 oz	95	0.000173 oz	0.998383 ozf/oz	0.000173 ozf
ε_{cf}	± 5.99×10⁻⁶ ozf/oz	95.45	2.995×10⁻⁶ ozf/oz	1.6 oz	4.792×10⁻⁶ ozf
ε_{res}	± 0.0025 ozf	100	0.00144 ozf	1	0.00144 ozf

The uncertainty in δ is computed by taking the root sum square of the component uncertainties.

$$u_\delta = \sqrt{(0.000173\ \text{ozf})^2 + \left(4.792 \times 10^{-6}\ \text{ozf}\right)^2 + (0.00144\ \text{ozf})^2}$$

$$= \sqrt{2.1036 \times 10^{-6}\ \text{ozf}^2} = 0.00145\ \text{ozf}$$

The pareto chart, shown in Figure F-1, indicates that the force gauge digital resolution is the largest contributor to the combined uncertainty. The uncertainties due to the weight standards and conversion factor provide much less contribution to the combined uncertainty.

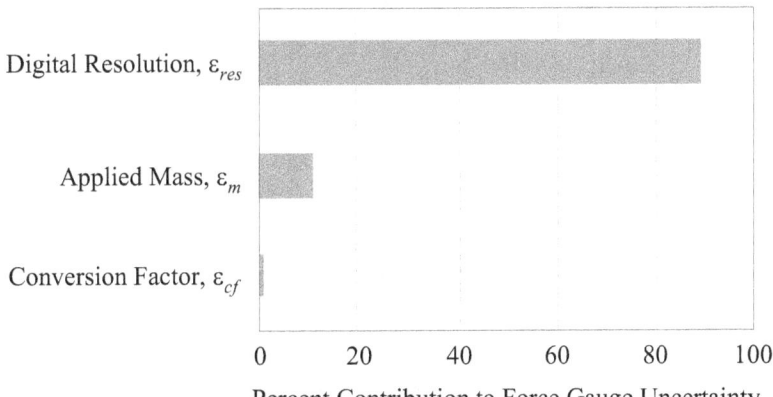

Figure F-1. Pareto Chart for Force Gauge Calibration Uncertainty

The Welch-Satterthwaite formula is used to compute the degrees of freedom for u_δ.

$$\nu_{u_\delta} = \frac{u_\delta^4}{\dfrac{\left(c_m u_{\varepsilon_m}\right)^4}{\nu_{u_{\varepsilon_m}}} + \dfrac{\left(c_{cf} u_{\varepsilon_{cf}}\right)^4}{\nu_{u_{\varepsilon_{cf}}}} + \dfrac{u_{\varepsilon_{res}}^4}{\nu_{u_{\varepsilon_{res}}}}} = \frac{u_\delta^4}{\dfrac{\left(c_m u_{\varepsilon_m}\right)^4}{\infty} + \dfrac{\left(c_{cf} u_{\varepsilon_{cf}}\right)^4}{\infty} + \dfrac{u_{\varepsilon_{res}}^4}{\infty}} \quad \text{(F-15)}$$

The degrees of freedom for all of the process uncertainties are infinite, so the combined uncertainty also has infinite degrees of freedom.

The uncertainty estimates for the tension mode and compression mode calibration data are listed in Tables F-6 and F-7, respectively.

Table F-6. Uncertainty Estimates for Tension Mode Calibration Data

Applied Mass (oz)	Applied Force (ozf)	Force Gauge Reading Average (ozf)	Measured Difference δ (ozf)	Standard Uncertainty u_δ (ozf)	Degrees of Freedom	Force Gauge Specification Limits (ozf)
0.0000	0.0000	0.000	0.0000	0.00144	∞	$\pm\,0.017$
1.6000	1.5974	1.598	0.0001	0.00145	∞	$\pm\,0.017$
3.2000	3.1948	3.195	0.0002	0.00148	∞	$\pm\,0.017$
4.8000	4.7922	4.793	0.0003	0.00153	∞	$\pm\,0.017$
6.4000	6.3897	6.388	-0.0022	0.00160	∞	$\pm\,0.017$
8.0000	7.9871	7.985	-0.0021	0.00167	∞	$\pm\,0.017$

Table F-7. Uncertainty Estimates for Compression Mode Calibration Data

Applied Mass (oz)	Applied Force (ozf)	Force Gauge Reading Average (ozf)	Measured Difference δ (ozf)	Standard Uncertainty u_δ (ozf)	Degrees of Freedom	Force Gauge Specification Limits (ozf)
0.0000	0.0000	0.000	0.0000	0.00144	∞	$\pm\,0.017$
1.6000	1.5974	1.595	-0.0024	0.00145	∞	$\pm\,0.017$
3.2000	3.1948	3.195	0.0002	0.00148	∞	$\pm\,0.017$
4.8000	4.7922	4.790	-0.0022	0.00153	∞	$\pm\,0.017$
6.4000	6.3897	6.385	-0.0047	0.00160	∞	$\pm\,0.017$
8.0000	7.9871	7.985	-0.0021	0.00167	∞	$\pm\,0.017$

F.3   In-tolerance Probability

As previously discussed, the probability that the UUT bias is in-tolerance is based on the calibration result and its associated uncertainty. The largest value of δ is -0.0047 oz-force with an associated uncertainty of 0.00160 oz-force. This value is an estimate of the bias, $e_{UUT,b}$, in the force gauge reading for an applied force of 6.3897 oz-force at the time of calibration.

Figure 2 shows the $\varepsilon_{UUT,b}$ probability distribution for the population of Chatillon Model DGGS-250G force gauges. The spread of the distribution is based on the manufacturer specified tolerance limits of ± 0.017 oz-force. The calibration result, $\delta = -0.0047\,\text{ozf}$, is depicted along with black bars showing 95% confidence limits computed from

$$\delta \pm t_{\alpha/2,\nu} \times u_\delta \qquad\qquad (\text{F-16})$$

where $t_{\alpha/2,\nu}$ is the t-statistic, $\alpha = p/2$, p is the confidence level and ν is the degrees of freedom for u_δ. For a 95% confidence level, $t_{0.025,\infty} = 1.9600$ and the confidence limits are computed to be

$$-0.0047\,\text{ozf} \pm 1.96 \times 0.00160\,\text{ozf} \ \text{ or } \ -0.0047\,\text{ozf} \pm 0.0031\,\text{ozf}.$$

This means that, while the value of $\varepsilon_{UUT,b}$ is unknown, there is a 95% confidence level that it is contained within the limits of $-0.0047\,\text{ozf} \pm 0.0031\,\text{ozf}$.

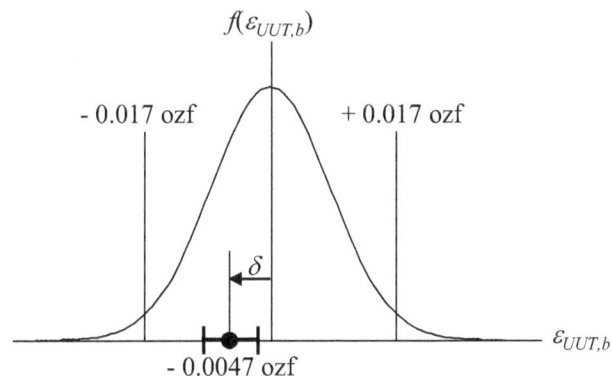

Figure F-2. Force Gauge Bias Distribution

The probability that $\varepsilon_{UUT,b}$ falls outside of the manufacturer specification limits is basically zero. So, the UUT can be considered to be in-tolerance over the calibrated force range.

APPENDIX G – SPECTRUM ANALYZER ANALYSIS EXAMPLE

The purpose of this analysis is to estimate and report the uncertainty in the minimum and maximum values of the relative flatness error of a spectrum analyzer that is calibrated over a frequency range of 50 MHz to 3.0 GHz.

G.1   Measurement Process Overview

The frequency response performance parameter of an Agilent E4440A Spectrum Analyzer is calibrated using an Agilent 5701B Signal Generator, Agilent 438A Power Meter, Agilent 11667B Power Divider, and Agilent 8485A Power Sensor, as shown in Figure G-1.

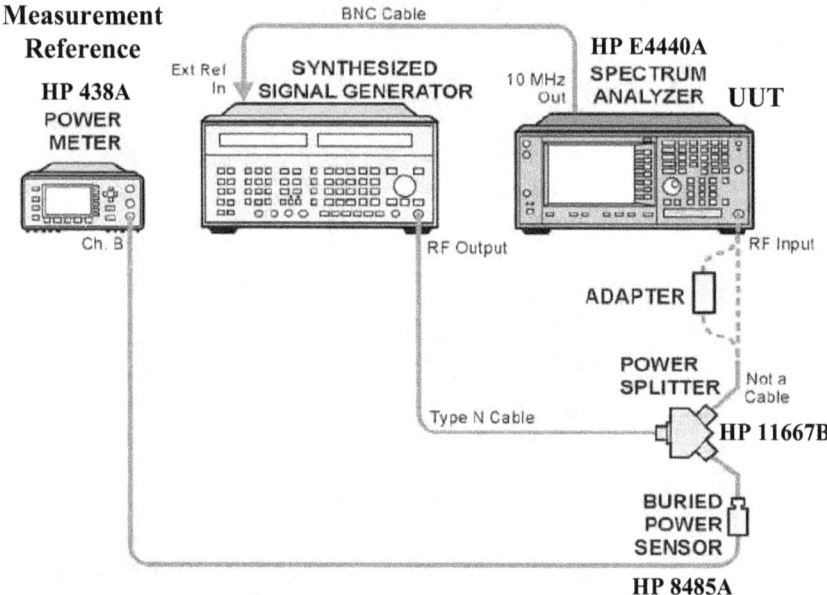

Figure G-1. Spectrum Analyzer Calibration Setup

The maximum frequency response error for the unit under test (UUT) spectrum analyzer is specified relative to the frequency response at 50 MHz.   As shown in Table 1, the maximum relative error varies for different frequency ranges. Agilent reports that these maximum relative error limits correspond to a 10 dB input attenuation (i.e., - 10 dB) and 20 °C to 30 °C environmental operating temperature range.

Table G-1. Frequency Response Specifications for Agilent E4440A[88]

Frequency Range	Max. Error Relative to 50 MHz Response
3 Hz to 3.0 GHz	± 0.38 dB
3.0 GHz to 6.6 GHz[b]	± 1.50 dB
6.6 GHz to 13.2 GHz[b]	± 2.00 dB
13.2 GHz to 22.0 GHz[b]	± 2.00 dB
22.0 GHz to 26.5 GHz[b]	± 2.50 dB

---

[88] Specifications Guide for PSA Series Spectrum Analyzers, Manufacturing Part Number E4440-90606, Printed in USA April 2009, Agilent Technologies, Inc.

b. Preselector centering applied.

The frequency response of the UUT is essentially an amplitude flatness specification for a given frequency range. The flatness error is defined as[89]

$$\delta_{flat} = A_{UUT} - A_{PM_B} \tag{G-1}$$

where

A_{UUT} = UUT amplitude measurement at calibration frequency

A_{PM_B} = Power meter amplitude measurement at calibration frequency

The calibration procedure calls for a nominal power input of -10 dBm (0.1 mW) to be supplied to the UUT.[90] As shown in Figure G-1, this is achieved by splitting the signal generator power output to the UUT and the power sensor/power meter. The flatness error is then computed at selected frequencies within a range (e.g., 3 Hz to 3 GHz).

The power splitter is initially characterized using an Agilent 8482A as a reference power sensor, as shown in Figure G-2. The reference power sensor is connected to Channel A of the power meter and the Agilent 8485A sensor power is connected to Channel B.

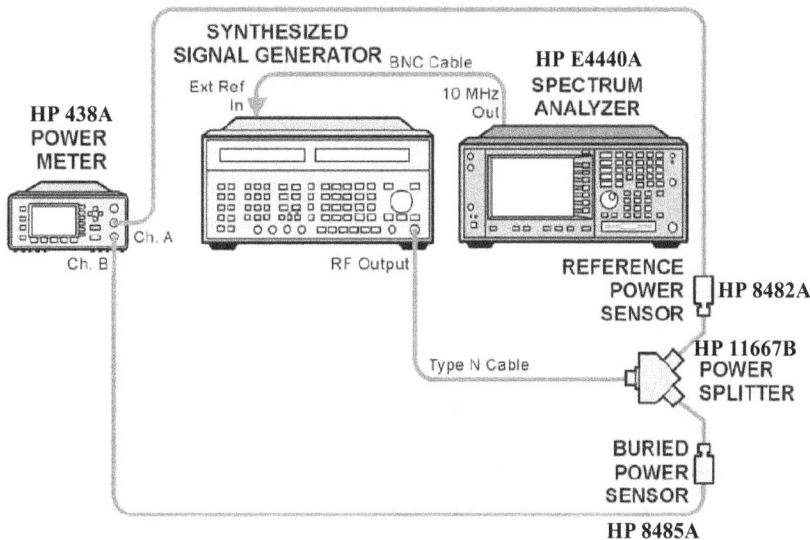

Figure G-2. Power Splitter Characterization Setup

The splitter tracking error is computed from equation (G-2).

$$\delta_{split} = A_{PM_A} - A_{PM_B} \tag{G-2}$$

where

---

[89] Instrument Messages and Functional Tests for PSA Series Spectrum Analyzers and ESA Series Spectrum Analyzers, Manufacturing Part Number E4440-90619, Printed in USA June 2008, Agilent Technologies, Inc.

[90] Need reference information for Agilent E4440A Spectrum Analyzer Performance Verification Tests Frequency Response.

$$A_{PM_A} = \text{Power meter Channel A measurement}$$

$$A_{PM_B} = \text{Power meter Channel B measurement}$$

The flatness error equation is then modified to account for the splitter tracker error.

$$\delta_{flat} = A_{UUT} - A_{PM_B} - \delta_{split} \qquad (\text{G-3})$$

The relative flatness error at a given test or calibration frequency is defined as

$$\delta_{flat_{rel}} = \delta_{flat} - \delta_{flat_{50MHz}} \qquad (\text{G-4})$$

where

$$\delta_{flat_{50MHz}} = \text{flatness error at 50 MHz frequency}$$

The minimum and maximum values of $\delta_{flat_{rel}}$ are recorded for the calibration frequency range and compared to the frequency response specification limits.[91] If either value of $\delta_{flat_{rel}}$ falls outside of these limits, then the UUT is typically deemed to be out-of-tolerance (OOT) or noncompliant. However, errors in the calibration process can result in an incorrect OOT assessment (false-reject) or in-tolerance assessment (false-accept).

The relationship between the calibration result, $\delta_{flat_{rel}}$, and the true UUT bias, $e_{UUT,b}$, is generally expressed as

$$\delta_{flat_{rel}} = e_{UUT,b} + \varepsilon_{cal} \qquad (\text{G-5})$$

where ε_{cal} is the calibration error.

The probability that the UUT frequency response parameter is in-tolerance is based on the calibration result and its associated uncertainty. Therefore, all relevant calibration error sources must be identified and combined in a way that yields viable uncertainty estimates.

G.2 Uncertainty Analysis Procedure

The purpose of this analysis is to estimate and report the total uncertainty in the minimum and maximum values of $\delta_{flat_{rel}}$ observed for a calibration frequency range of 50 MHz to 3.0 GHz. The calibration results are summarized in Table G-2.

Table G-2. Relative Flatness Error for 50 MHz to 3 GHz Range

	$\delta_{flat_{rel}}$ (dB)	Specification Limits (dB)	$\delta_{flat_{rel}}$ [a] (µW)	Specification Limits (µW)
Minimum	0.13	± 0.38	3.04	+ 9.14

---

[91] Since the tolerance limits constitute the maximum permissible deviation or difference, they should be expressed in units that are consistent with those measured during calibration.

				- 8.38
Maximum	0.32	± 0.38	7.65	+ 9.14 - 8.38

a. Input Power = 0.1 mW

The uncertainty in $\delta_{flat_{rel}}$ is determined by applying the variance operator to equation (G-5) and taking the square root.

$$u_{\delta_{flat_{rel}}} = \sqrt{\mathrm{var}\left(\delta_{flat_{rel}}\right)} = \sqrt{\mathrm{var}\left(e_{UUT,b} + \varepsilon_{cal}\right)}$$
$$= \sqrt{\mathrm{var}\left(\varepsilon_{cal}\right)} \tag{G-6}$$

Given the equipment and procedures used, the calibration error equation is

$$\varepsilon_{cal} = \varepsilon_{UUT,res} + \varepsilon_{\delta_{split}} + \varepsilon_{PM} + \varepsilon_{PS_B} \tag{G-7}$$

where

$$\varepsilon_{\delta_{split}} = \varepsilon_{PS_A} + \varepsilon_{PS_B} + 2\varepsilon_{PM} \tag{G-8}$$

$$\varepsilon_{PM} = \varepsilon_{PM,b} + \varepsilon_{PM,res} \tag{G-9}$$

and

$\varepsilon_{UUT,res}$	=	UUT resolution error
ε_{PS_A}	=	Channel A power sensor bias
ε_{PS_B}	=	Channel B power sensor bias
$\varepsilon_{PM,b}$	=	Power meter bias
$\varepsilon_{PM,res}$	=	Power meter resolution error

Note: During the power splitter characterization, the two power sensors are connected to different power meter channels. Consequently, the power meter error contributes to the error in δ_{split} via the Channel A power and Channel B power measurements, as depicted in equation (G-8).

Substituting equations (G-8) and (G-9) into (G-7), the calibration error equation becomes

$$\varepsilon_{cal} = \varepsilon_{UUT,res} + \varepsilon_{PS_A} + 2\varepsilon_{PS_B} + 3\varepsilon_{PM,b} + 3\varepsilon_{PM,res} \tag{G-10}$$

Brief descriptions of the calibration process errors are provided in the following subsections.

G.2.1 UUT Resolution Error ($\varepsilon_{UUT,res}$)

The Agilent E4440A Spectrum Analyzer has a digital display resolution of 0.01 dB or $\leq$ 1% of the input signal level.[92] Therefore, the resolution error limits in dB units are ± 0.005 dB (i.e., ±

---

[92] Specifications Guide for PSA Series Spectrum Analyzers, Manufacturing Part Number E4440-90606, Printed in USA April 2009, Agilent Technologies, Inc.

half the resolution). The resolution error limits in mW units are computed to be

$$= \pm \frac{1}{100} \times \frac{0.1\,\text{mW}}{2} = \pm 0.0005\,\text{mW} = \pm 0.5\,\mu\text{W}$$

These limits represent 100% containment limits for a uniformly distributed error.

G.2.2 Channel A Power Sensor Bias (ε_{PS_A})

During the power splitter characterization, the HP8482A power sensor is connected to the power meter Channel A. The most recent calibration data sheet[93] for this power sensor states an expanded uncertainty of $\pm$ 0.99 % of the sensed power for a frequency range of 50 MHz to 3 GHz. The expanded uncertainty limits are assumed to represent a coverage factor of $k = 2$.

G.2.3 Channel B Power Sensor Bias (ε_{PS_B})

The HP8485A power sensor is connected to the power meter Channel B for both the power splitter characterization and the UUT frequency response calibration. The most recent calibration data sheet[94] for this power sensor states an expanded uncertainty of $\pm$ 1.75 % of the sensed power for a frequency range or 50 MHz to 3 GHz. The expanded uncertainty limits are assumed to represent a coverage factor of $k = 2$.

G.2.4 Power Meter Error ($\varepsilon_{PM,b}$ and $\varepsilon_{PM,res}$)

The Agilent 438A power meter is a microprocessor controlled dual channel meter that is used in conjunction with an Agilent 8480 series power sensor to measure power ranging from -70 to +44 dBm (100 pW to 25 W) for a frequency range of 100 kHz to 26.5 GHz .

The accuracy limits for the 438A power meter are specified to be $\pm$ 0.02 dB (single channel mode). The accuracy limits in μW units are computed to be

$$= \pm \left(10^{(0.02/10)} \times 0.1\,\text{mW} - 0.1\,\text{mW} \right)$$
$$= \pm \left(1.00462 \times 0.1\,\text{mW} - 0.1\,\text{mW} \right)$$
$$= \pm 0.000462\,\text{mW} = \pm 0.462\,\mu\text{W}$$

The accuracy limits are assumed to represent 95% confidence limits for a normally distributed error.

The digital display resolution is specified to be 0.1% full scale. The resolution error limits for the 0.01 to 0.1mW range are computed to be

$$= \pm \frac{0.1}{100} \times \frac{0.1\,\text{mW}}{2}$$
$$= \pm 0.00005\,\text{mW} = \pm 0.05\,\mu\text{W}$$

---

[93] Wyle Laboratories Calibration Data Sheet Number M78587-20May08, Model 8482A, Serial Number US37294071.

[94] Wyle Report of Test Number 7.54389.04, Model 8485A, Serial Number 2703A05070, September 5, 2008.

The resolution limits represent 100% containment limits for a uniformly distributed error.

Applying the variance operator to equation (G-10), gives

$$
\begin{aligned}
\operatorname{var}\left(\varepsilon_{cal}\right) &= \operatorname{var}\left(\varepsilon_{UUT,res} + \varepsilon_{PS_A} + 2\varepsilon_{PS_B} + 3\varepsilon_{PM,b} + 3\varepsilon_{PM,res}\right) \\
&= \operatorname{var}\left(\varepsilon_{UUT,res}\right) + \operatorname{var}\left(\varepsilon_{PS_A}\right) + 4\operatorname{var}\left(\varepsilon_{PS_B}\right) + 9\operatorname{var}\left(\varepsilon_{PM,b}\right) + 9\operatorname{var}\left(\varepsilon_{PM,res}\right) \\
&\quad + 2\operatorname{cov}\left(\varepsilon_{UUT,res},\varepsilon_{PS_A}\right) + 4\operatorname{cov}\left(\varepsilon_{UUT,res},\varepsilon_{PS_B}\right) + 6\operatorname{cov}\left(\varepsilon_{UUT,res},\varepsilon_{PM,b}\right) \\
&\quad + 6\operatorname{cov}\left(\varepsilon_{UUT,res},\varepsilon_{PM,res}\right) + 4\operatorname{cov}\left(\varepsilon_{PS_A},\varepsilon_{PS_B}\right) + 6\operatorname{cov}\left(\varepsilon_{PS_A},\varepsilon_{PM,b}\right) \\
&\quad + 6\operatorname{cov}\left(\varepsilon_{PS_A},\varepsilon_{PM,res}\right) + 12\operatorname{cov}\left(\varepsilon_{PS_B},\varepsilon_{PM,b}\right) + 12\operatorname{cov}\left(\varepsilon_{PS_B},\varepsilon_{PM,res}\right) \\
&\quad + 18\operatorname{cov}\left(\varepsilon_{PM,b},\varepsilon_{PM,res}\right)
\end{aligned}
\tag{G-11}
$$

where the cov() terms account for the covariance between pairs of error sources. Covariance is a statistical assessment of the mutual dependence of the errors. The covariance terms can have inconvenient physical dimensions, so the correlation coefficient is often used instead. For example, the correlation coefficient for $\varepsilon_{UUT,res}$ and ε_{PS_A} is defined as

$$
\rho_{\varepsilon_{UUT,res}\varepsilon_{PS_A}} = \frac{\operatorname{cov}\left(\varepsilon_{UUT,res},\varepsilon_{PS_A}\right)}{u_{\varepsilon_{UUT,res}}\,u_{\varepsilon_{PS_A}}}
$$

where $u_{\varepsilon_{UUT,res}}$ and $u_{\varepsilon_{PS_A}}$ are the uncertainties in $\varepsilon_{UUT,res}$ and ε_{PS_A}, respectively. Therefore, equation (G-11) can be expressed as

$$
\begin{aligned}
\operatorname{var}\left(\varepsilon_{cal}\right) &= \operatorname{var}\left(\varepsilon_{UUT,res}\right) + \operatorname{var}\left(\varepsilon_{PS_A}\right) + 4\operatorname{var}\left(\varepsilon_{PS_B}\right) + 9\operatorname{var}\left(\varepsilon_{PM,b}\right) + 9\operatorname{var}\left(\varepsilon_{PM,res}\right) \\
&\quad + 2\rho_{\varepsilon_{UUT,res},\varepsilon_{PS_A}} u_{\varepsilon_{UUT,res}} u_{\varepsilon_{PS_A}} + 4\rho_{\varepsilon_{UUT,res},\varepsilon_{PS_B}} u_{\varepsilon_{UUT,res}} u_{\varepsilon_{PS_B}} \\
&\quad + 6\rho_{\varepsilon_{UUT,res},\varepsilon_{PM,b}} u_{\varepsilon_{UUT,res}} u_{\varepsilon_{PM,b}} + 6\rho_{\varepsilon_{UUT,res},\varepsilon_{PM,res}} u_{\varepsilon_{UUT,res}} u_{\varepsilon_{PM,res}} \\
&\quad + 4\rho_{\varepsilon_{PS_A},\varepsilon_{PS_B}} u_{\varepsilon_{PS_A}} u_{\varepsilon_{PS_B}} + 6\rho_{\varepsilon_{PS_A},\varepsilon_{PM,b}} u_{\varepsilon_{PS_A}} u_{\varepsilon_{PM,b}} \\
&\quad + 6\rho_{\varepsilon_{PS_A},\varepsilon_{PM,res}} u_{\varepsilon_{PS_A}} u_{\varepsilon_{PM,res}} + 12\rho_{\varepsilon_{PS_B},\varepsilon_{PM,b}} u_{\varepsilon_{PS_B}} u_{\varepsilon_{PM,b}} \\
&\quad + 12\rho_{\varepsilon_{PS_B},\varepsilon_{PM,res}} u_{\varepsilon_{PS_B}} u_{\varepsilon_{PM,res}} + 18\rho_{\varepsilon_{PM,b},\varepsilon_{PM,res}} u_{\varepsilon_{PM,b}} u_{\varepsilon_{PM,res}}
\end{aligned}
\tag{G-12}
$$

There are no correlations between error sources, so equation (G-12) can be simplified to

$$
\operatorname{var}\left(\varepsilon_{cal}\right) = \operatorname{var}\left(\varepsilon_{UUT,res}\right) + \operatorname{var}\left(\varepsilon_{PS_A}\right) + 4\operatorname{var}\left(\varepsilon_{PS_B}\right) + 9\operatorname{var}\left(\varepsilon_{PM,b}\right) + 9\operatorname{var}\left(\varepsilon_{PM,res}\right)
\tag{G-13}
$$

The variance terms in equation (G-13) are equivalent to the square of the uncertainty in the corresponding error (e.g., $\operatorname{var}(\varepsilon_{PS_A}) = u_{\varepsilon_{PS_A}}^2$). So, the uncertainty equation for $\delta_{flat_{rel}}$ can be rewritten in terms of the individual measurement process uncertainties and their associated sensitivity coefficients.

$$
\begin{aligned}
u_{\delta_{flat_{rel}}} &= \sqrt{\text{var}\left(\varepsilon_{cal}\right)} \\
&= \sqrt{u^2_{\varepsilon_{UUT,res}} + u^2_{\varepsilon_{PS_A}} + 4u^2_{\varepsilon_{PS_B}} + 9u^2_{\varepsilon_{PM,b}} + 9u^2_{\varepsilon_{PM,res}}}
\end{aligned}
\tag{G-14}
$$

The spectrum analyzer digital resolution uncertainty is estimated using the $\pm\,0.5\;\mu$W error limits, the inverse uniform distribution function and a 1.00 containment probability (100% confidence level).

$$
u_{\varepsilon_{UUT,res}} = \frac{0.5\,\mu\text{W}}{\sqrt{3}} = \frac{0.5\,\mu\text{W}}{1.732} = 0.29\,\mu\text{W}
$$

The Channel A power sensor bias uncertainty is estimated using the expanded uncertainty of $\pm$ 0.99% of the sensed power, a sensed power of 0.1 mW and $k = 2$ coverage factor.

$$
u_{\varepsilon_{PS_A}} = \frac{\dfrac{0.99}{100} \times 0.1\,\text{mW}}{2} = \frac{0.00099\,\text{mW}}{2} = \frac{0.99\,\mu\text{W}}{2} = 0.495\,\mu\text{W}
$$

The Channel B power sensor bias uncertainty is estimated using the $\pm$ 1.75% expanded uncertainty, the sensed power = 0.1 mW and $k = 2$ coverage factor.

$$
u_{\varepsilon_{PS_B}} = \frac{\dfrac{1.75}{100} \times 0.1\,\text{mW}}{2} = \frac{0.00175\,\text{mW}}{2} = \frac{1.75\,\mu\text{W}}{2} = 0.875\,\mu\text{W}
$$

The power meter bias uncertainty is estimated using the $\pm\,0.462\;\mu$W error limits, the inverse normal distribution function and a 0.95 containment probability (95% confidence level).

$$
u_{\varepsilon_{PM,b}} = \frac{0.462\,\mu\text{W}}{\Phi^{-1}\left(\dfrac{1+0.95}{2}\right)} = \frac{0.462\,\mu\text{W}}{1.9600} = 0.236\,\mu\text{W}
$$

The power meter digital resolution uncertainty is estimated using the $\pm\,0.05\;\mu$W error limits, the inverse uniform distribution function and a 1.00 containment probability (100% confidence level).

$$
u_{\varepsilon_{PM,res}} = \frac{0.05\,\mu\text{W}}{\sqrt{3}} = \frac{0.05\,\mu\text{W}}{1.732} = 0.029\,\mu\text{W}
$$

The estimated uncertainties and sensitivity coefficients for each error source are summarized in Table G-3. The component uncertainties are the product of the standard uncertainty and the sensitivity coefficient.

217

Table G-3. Estimated Uncertainties for Agilent E4440A Frequency Response Calibration

Error Source	± Error Limits	Conf. Level	Standard Uncertainty	Deg. Freedom	Sensitivity Coefficient	Component Uncertainty
$\varepsilon_{UUT,res}$	± 0.5 μW	100	0.29 μW	∞	1	0.29 μW
ε_{PS_A}	± 0.99 μW	95.45	0.495 μW	∞	1	0.495 μW
ε_{PS_B}	± 1.75 μW	95.45	0.875 μW	∞	2	1.75 μW
$\varepsilon_{PM,b}$	± 0.462 μW	95	0.236 μW	∞	3	0.708 μW
$\varepsilon_{PM,res}$	± 0.05 μW	100	0.029 μW	∞	3	0.087 μW

The uncertainty in $\delta_{flat_{rel}}$ is computed by taking the root sum square of the component uncertainties.

$$u_{\delta_{flat_{rel}}} = \sqrt{(0.29\,\mu W)^2 + (0.495\,\mu W)^2 + (1.75\,\mu W)^2 + (0.708\,\mu W)^2 + (0.087\,\mu W)^2}$$
$$= \sqrt{0.084 + 0.245 + 3.063 + 0.501 + 0.0076}\,\mu W$$
$$= \sqrt{3.900}\,\mu W$$
$$= 1.975\,\mu W$$

The pareto chart, shown in Figure G-3, indicates that the Channel B power sensor bias uncertainty is the largest contributor to the uncertainty in the UUT relative flatness error.

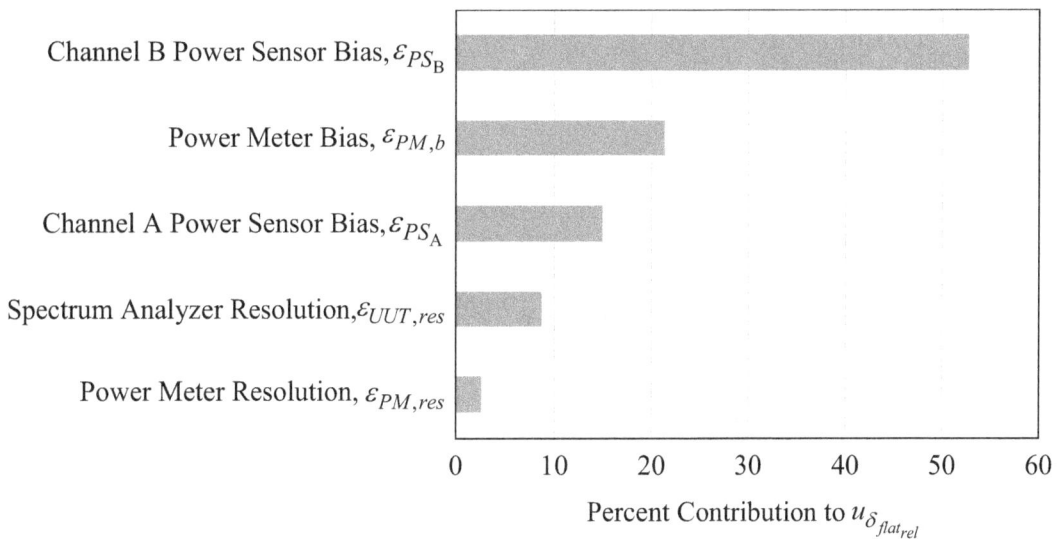

Figure G-3. Pareto Chart for UUT Frequency Response Calibration

The Welch-Satterthwaite formula given in equation (G-15) is used to compute the degrees of freedom for $u_{\delta_{flat_{rel}}}$. The degrees of freedom for all of the process uncertainties are infinite, so the uncertainty in the relative flatness error also has infinite degrees of freedom.

$$v_{u_{\delta_{flat_{rel}}}} = \frac{u_{\delta_{flat_{rel}}}^4}{\dfrac{\left(u_{\varepsilon_{UUT,res}}\right)^4}{\infty} + \dfrac{\left(u_{\varepsilon_{PS_A}}\right)^4}{\infty} + \dfrac{\left(2u_{\varepsilon_{PS_B}}\right)^4}{\infty} + \dfrac{\left(3u_{\varepsilon_{PM,b}}\right)^4}{\infty} + \dfrac{\left(3u_{\varepsilon_{PM,res}}\right)^4}{\infty}} \qquad \text{(G-15)}$$

The uncertainties for the minimum and maximum values of $u_{\delta_{flat_{rel}}}$ are summarized in Table G-4.

Table G-4. Estimated Uncertainties for Relative Flatness Error (50 MHz to 3 GHz)

	$\delta_{flat_{rel}}$ (dB)	$\delta_{flat_{rel}}$ (μW)	$u_{\delta_{flat_{rel}}}$ (μW)	Specification Limits (μW)
Minimum	0.13	3.04	1.975	+ 9.14 - 8.38
Maximum	0.32	7.65	1.975	+ 9.14 - 8.38

G.3 In-tolerance Probability

As previously discussed, the probability that the bias in the UUT frequency response parameter, $\varepsilon_{UUT,b}$, is in-tolerance is based on the calibration result and its associated uncertainty. The maximum relative flatness error was determined to be 7.65 μW. This value is an estimate of $e_{UUT,b}$, at the time of calibration.

Figure G-4 shows the probability distribution for $e_{UUT,b}$. The spread of the distribution is based on the specified tolerance limits of + 9.144 μW and − 8.378 μW for Agilent E4440A Spectrum Analyzers.

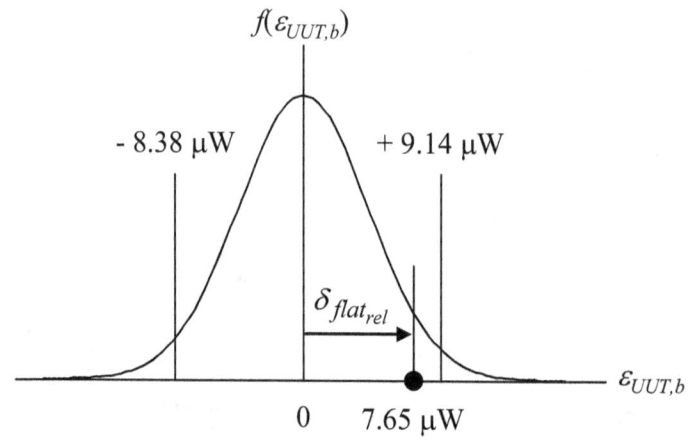

Figure G-4. Bias Distribution for UUT Frequency Response

Given the maximum value of $\delta_{flat_{rel}}$ observed during calibration, it appears that the UUT frequency response parameter is in-tolerance. However, because of the uncertainty in $\delta_{flat_{rel}}$, the actual bias in the UUT frequency response parameter, $\varepsilon_{UUT,b}$, may be larger or smaller than 7.65 μW.

The confidence limits for $\varepsilon_{UUT,b}$ can be expressed as

$$\delta_{flat_{rel}} \pm t_{\alpha/2,v} \times u_{\delta_{flat_{rel}}} \tag{G-16}$$

where $t_{\alpha/2,v}$ is the t-statistic, $\alpha = p/2$, p is the confidence level and v is the degrees of freedom for $u_{\delta_{flat_{rel}}}$.

For a 95% confidence level, $t_{0.025,\infty} = 1.9600$ and the confidence limits for $e_{UUT,b}$ are computed to be

$$7.65\ \mu W \pm 1.96 \times 1.975\ \mu W \text{ or } 7.65\ \mu W \pm 3.87\ \mu W.$$

This means that, while the value of $\varepsilon_{UUT,b}$ is unknown, there is a 95% confidence that it is contained within the limits of $7.65\ \mu W \pm 3.87\ \mu W$.

Figure G-5 shows the probability distribution for $\varepsilon_{UUT,b}$ given the calibration result $\delta_{flat_{rel}} = 7.65$ μW. The black bar depicts the $\pm\ 3.87\ \mu W$ confidence limits. Given the relatively large uncertainty in $\delta_{flat_{rel}}$, the in-tolerance probability of $\varepsilon_{UUT,b}$ appears to be significantly reduced.

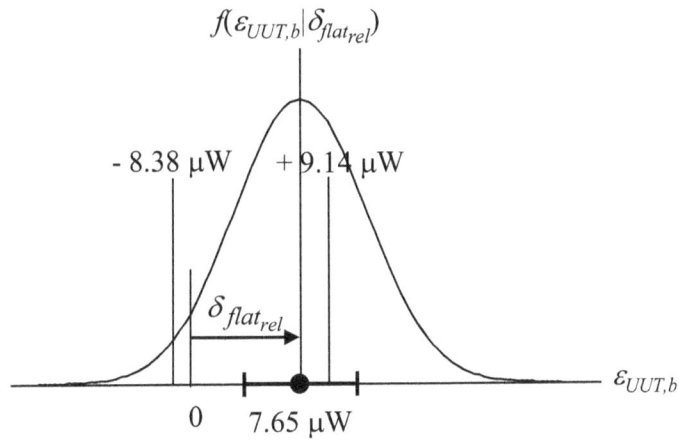

Figure G-5. OOT Probability of UUT Frequency Response

Bayesian analysis methods are employed to estimate the true value of $\varepsilon_{UUT,b}$ and compute the in-tolerance probability based on *a priori* knowledge and on measurement results obtained during calibration.[95]

Prior to calibration, the uncertainty, $u_{\varepsilon_{UUT,b}}$, is estimated from the in-inverse probability distribution for $\varepsilon_{UUT,b}$, the specification limits and the associated *a priori* in-tolerance probability. In this analysis, two underlying assumptions are employed:

---

[95] An in-depth coverage of the methods and principles used to compute in-tolerance probability are provided in NASA *Measurement Quality Assurance* Handbook, Annex 4 – *Estimation and Evaluation of Measurement Decision Risk.*

1. $\varepsilon_{UUT,b}$ is assumed to be normally distributed.

2. The asymmetric tolerance limits $+L_1$ and $-L_2$ are *a prior* 95% confidence limits.

Since $L_1 \neq L_2$, the value of $u_{\varepsilon_{UUT,b}}$ is computed by solving equation (G-17) through numerical iteration.

$$P_a(in) = \Phi\left(\frac{L_1}{u_{\varepsilon_{UUT,b}}}\right) + \Phi\left(\frac{L_2}{u_{\varepsilon_{UUT,b}}}\right) - 1 \qquad (G\text{-}17)$$

where $P_a(in)$ is the *a priori* in-tolerance probability of 95% and Φ is the normal distribution function.

A value of $u_{\varepsilon_{UUT,b}} = 4.45\ \mu W$ was computed off-line using an uncertainty analysis software program. This bias uncertainty estimate is equivalent to the standard deviation of the probability distribution for the population shown in Figure G-4.

After calibration, the values of $u_{\varepsilon_{UUT,b}}$, $\delta_{flat_{rel}}$ and $u_{\delta_{flat_{rel}}}$ are used to estimate the true value of $\varepsilon_{UUT,b}$. This value is denoted β and computed from equation (G-18).

$$\beta = \frac{u_{\varepsilon_{UUT,b}}^2}{u_{\varepsilon_{UUT,b}}^2 + u_{\delta_{flat_{rel}}}^2} \times \delta_{flat_{rel}} \qquad (G\text{-}18)$$

The Bayesian estimate, β, will be less than or equal to the calibration result, $\delta_{flat_{rel}}$. For example, if the values of $u_{\varepsilon_{UUT,b}}$ and $u_{\delta_{flat_{rel}}}$ are equal, then $\beta = \delta_{flat_{rel}}/2$. Conversely if $u_{\delta_{flat_{rel}}}$ is much smaller than $u_{\varepsilon_{UUT,b}}$, then $\beta \cong \delta_{flat_{rel}}$. From equation (G-18), the estimated true value of $\varepsilon_{UUT,b}$ is computed to be

$$\beta = \frac{(4.45\ \mu W)^2}{(4.45\ \mu W)^2 + (1.975\ \mu W)^2} \times 7.65\ \mu W$$

$$= \frac{19.80\ \mu W^2}{23.70\ \mu W^2} \times 7.65\ \mu W = 0.835 \times 7.65\ \mu W = 6.39\ \mu W.$$

This minor reduction in the UUT bias estimate reflects the fact that the calibration uncertainty is much smaller than the UUT bias uncertainty that is expected from the manufacturer specification limits. Therefore, the observed 7.65 μW deviation is considered to be mainly attributable to the UUT parameter bias.

The uncertainty in the Bayesian estimate β is computed from equation (G-19).

$$u_\beta = \frac{u_{\varepsilon_{UUT,b}}}{\sqrt{u_{\varepsilon_{UUT,b}}^2 + u_{\delta_{flat_{rel}}}^2}} \times u_{\delta_{flat_{rel}}} \qquad (G\text{-}19)$$

The uncertainty in β is computed to be

$$u_\beta = \frac{4.45\ \mu W}{\sqrt{(4.45\ \mu W)^2 + (1.975\ \mu W)^2}} \times 1.975\ \mu W$$

$$= \frac{4.45\ \mu W}{4.87\ \mu W} \times 1.975\ \mu W = 0.914 \times 1.975\ \mu W = 1.805\ \mu W.$$

Finally, the post-calibration in-tolerance probability for $\varepsilon_{UUT,b}$ is computed from equation (G-20).

$$P(in) = \Phi\left(\frac{L_1 + \beta}{u_\beta}\right) + \Phi\left(\frac{L_2 - \beta}{u_\beta}\right) - 1 \qquad (G\text{-}20)$$

The probability that the UUT frequency response (i.e., relative flatness error) is in-tolerance during calibration is computed to be

$$P(in) = \Phi\left(\frac{8.38\ \mu W + 6.39\ \mu W}{1.805\ \mu W}\right) + \Phi\left(\frac{9.14\ \mu W - 6.39\ \mu W}{1.805\ \mu W}\right) - 1$$

$$= \Phi\left(\frac{14.77}{1.805}\right) + \Phi\left(\frac{2.75}{1.805}\right) - 1 = \Phi(8.183) + \Phi(0.936) - 1$$

$$= 1.000 + 0.936 - 1 = 0.936 \text{ or } 93.6\%.$$

The resulting in-tolerance probability reflects the revised $\varepsilon_{UUT,b}$ estimate and its associated uncertainty. Figure G-6 shows the value of β with black bars that depict 95% confidence limits equal to $\pm\ 1.96 \times 1.805\ \mu W$ or $\pm\ 3.54\ \mu W$.

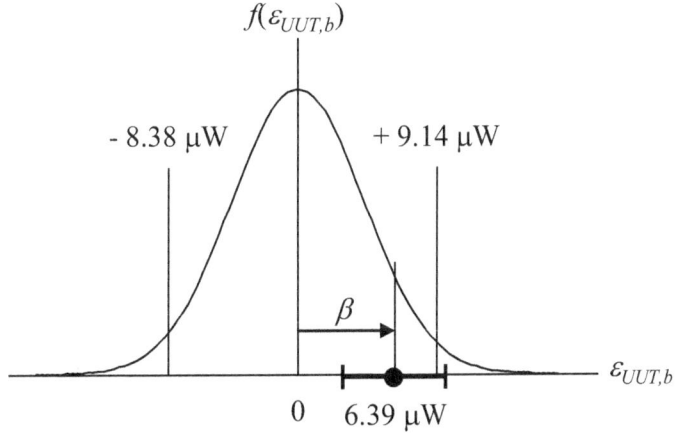

Figure G-6. Bayesian UUT Parameter Bias Estimate

APPENDIX H – ROTAMETER ANALYSIS EXAMPLE

The purpose of the rotameter uncertainty analysis is to

1. Estimate the uncertainties in gas flow rates used in the calibration of the rotameter.

2. Use these uncertainties in the establishment of the rotameter regression equation (i.e., calibration curve).

3. Estimate the uncertainties in flow rates predicted from the rotameter regression equation.

H.1   Measurement Process Overview

A Brooks model 1110 Series rotameter is calibrated with nitrogen gas using a Sierra Instruments Series 101 Cal-Bench as the measurement reference, as shown in Figure H-1. The temperature and pressure of the gas exiting the rotameter are measured with a Rosemount model 162C platinum resistance thermometer (PRT) and a Wallace & Tiernan FA-139 Precision Aneroid Barometer, respectively.

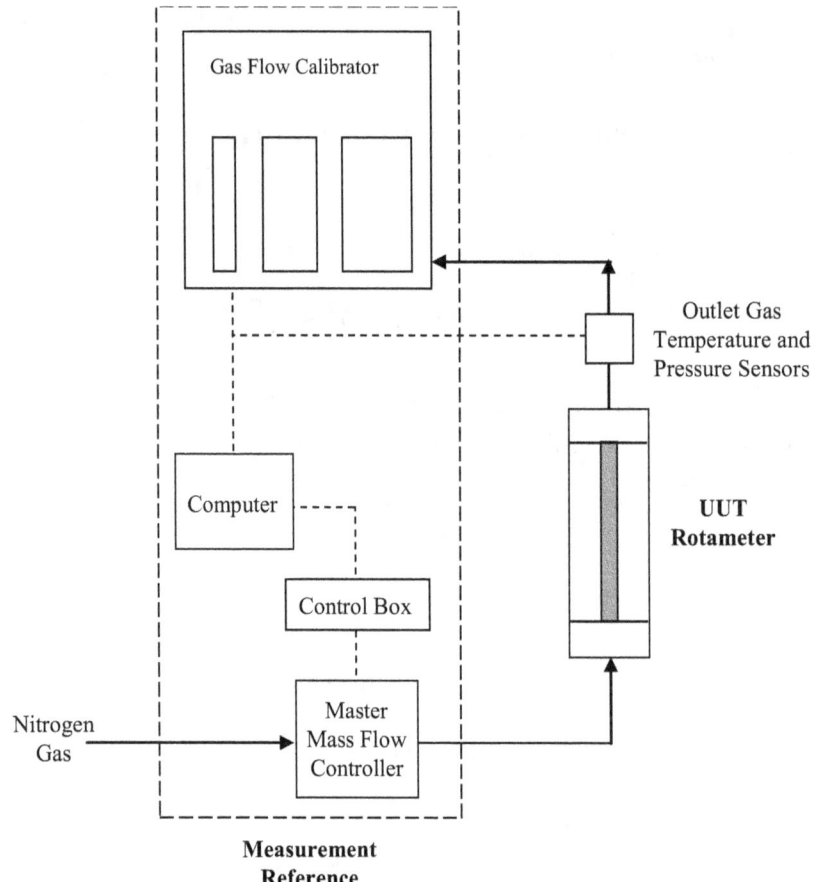

Figure H-1. Rotameter Calibration Setup

The rotameter scale readings at the center of the steel ball are collected for six flow rates. The measurement reference flow rates are corrected for the outlet gas pressure and temperature using equation (H-1).

223

$$R_{corr} = R_{ref} \times \left(\frac{P_{std}}{P_{out}} \times \frac{T_{out}}{T_{std}} \right)^{1/2} \qquad \text{(H-1)}$$

where

R_{ref} = measurement reference mass flow rate in standard cubic centimeters per minute (sccm)

P_{out} = outlet gas pressure, psia

T_{out} = outlet gas temperature, °R

P_{std} = standard pressure, 14.696 psia

T_{std} = standard temperature, 529.67 °R

The resulting calibration data are listed in Table H-1. The primary purpose of the calibration is to establish a third-order polynomial regression equation using the corrected flow rates and corresponding rotameter scale readings.

Table H-1. Rotameter Calibration Data (Steel Ball)96

UUT Rotameter Scale Reading	Outlet Gas Pressure (psia)	Outlet Gas Temperature (°C)	Measurement Reference Mass Flow Rate (sccm)	Corrected Flow Rate (ccm)
15.00	14.850	23.1	548.69	547.68
30.00	14.850	23.1	1356.75	1354.25
60.00	14.850	23.0	2937.12	2931.21
90.00	14.855	22.9	4551.52	4540.84
120.00	14.855	22.9	6227.96	6213.34
150.00	14.860	22.6	7881.43	7857.62

H.1.1 Regression Analysis

In regression analysis, a trend line is fit to the observed data. For example, a third-order regression equation is expressed as

$$\hat{y} = b_0 + b_1 x + b_2 x^2 + b_3 x^3 \qquad \text{(H-2)}$$

where $\hat{y}_i$ is the predicted value (e.g., flow rate), x is the corresponding independent variable (e.g., rotameter scale reading) and b_0, b_1, b_2 and b_3 are the regression coefficients. The regression coefficients are determined by minimizing the residual sum of squares (*RSS*).

$$RSS = \sum_{i=1}^{k} w_i \left(y_i - \hat{y}_i \right)^2 \qquad \text{(H-3)}$$

where

w_i = weighting factor

y_i = measured value

k = number of measured values used in regression analysis

---

[96] Wyle Laboratories Calibration Data Sheet, Metrology Number Z56093, 27Feb08

The weighting factor w_i is defined as

$$w_i = \left(\frac{\sigma}{\sigma_i}\right)^2 \tag{H-4}$$

where σ_i is the uncertainty in y_i and

$$\sigma^2 = \frac{k}{\displaystyle\sum_{i=1}^{k} \frac{1}{\sigma_i^2}} \tag{H-5}$$

Once the regression coefficients are determined, then the stage is set for predicting values of $\hat{y}_i$ given values of x. Each predicted value also has an associated uncertainty that must be estimated.

H.2   Uncertainty Analysis Procedure

To estimate the uncertainty in the corrected flow rates listed in Table H-1, all relevant measurement process errors must be identified and combined in an appropriate manner.

Given equation (H-1), the error equation for R_{corr} is

$$\varepsilon_{R_{corr}} = c_{R_{ref}}\varepsilon_{R_{ref}} + c_{P_{out}}\varepsilon_{P_{out}} + c_{T_{out}}\varepsilon_{T_{out}} \tag{H-6}$$

where

$\varepsilon_{R_{ref}}$ = error in the measurement reference flow rate

$\varepsilon_{P_{out}}$ = error in the outlet gas pressure measurement

$\varepsilon_{T_{out}}$ = error in the outlet gas temperature measurement

The coefficients in equation (H-6) are sensitivity coefficients that determine the relative contributions of the error sources to the overall error in the corrected flow rate.

The error in the measurement reference flow rate is comprised of errors due to the bias and resolution of the Sierra Instruments 101 Cal-Bench.

$$\varepsilon_{R_{ref}} = \varepsilon_{R_{ref,b}} + \varepsilon_{R_{ref,r}} \tag{H-7}$$

where

$\varepsilon_{R_{ref,b}}$ = measurement reference bias

$\varepsilon_{R_{ref,r}}$ = measurement reference resolution

Similarly, the error in the outlet gas pressure measurement is comprised of errors due to the bias and resolution of the Wallace & Tiernan FA-139 Precision Aneroid Barometer.

225

$$\varepsilon_{P_{out}} = \varepsilon_{P_{out,b}} + \varepsilon_{P_{out,r}} \tag{H-8}$$

where

$\varepsilon_{P_{out,b}}$ = barometer measurement bias

$\varepsilon_{P_{out,r}}$ = barometer resolution error

The error in the outlet gas temperature measurement is comprised of errors due to the bias and resolution of the Rosemount model 162C PRT.

$$\varepsilon_{T_{out}} = \varepsilon_{T_{out,b}} + \varepsilon_{T_{out,r}} \tag{H-9}$$

where

$\varepsilon_{T_{out,b}}$ = PRT measurement bias

$\varepsilon_{T_{out,r}}$ = PRT resolution error

Substituting equations (H-7) through (H-9) into equation (H-6), the error equation for the corrected flow rate can be expressed as

$$\varepsilon_{R_{corr}} = c_{R_{ref}}\left(\varepsilon_{R_{ref,b}} + \varepsilon_{R_{ref,r}}\right) + c_{P_{out}}\left(\varepsilon_{P_{out,b}} + \varepsilon_{P_{out,r}}\right) + c_{T_{out}}\left(\varepsilon_{T_{out,b}} + \varepsilon_{T_{out,r}}\right) \tag{H-10}$$

Brief descriptions of the measurement process errors are provided below.

H.2.1 Measurement Reference Flow Rate ($\varepsilon_{R_{ref,b}}$ and $\varepsilon_{R_{ref,r}}$)

The most recent calibration report[97] for the Sierra Instruments Series 101 Cal-Bench indicates an expanded uncertainty of $\pm 0.5\%$ of reading. The expanded uncertainty corresponds to coverage factor of $k = 2$. In this analysis, the measurement reference bias is assumed to follow a normal distribution.

The Series 101 Cal-Bench has a digital resolution of 0.001 sccm, so the resolution error limits are ± 0.0005 sccm (i.e., half the resolution). The digital resolution error is uniformly distributed with error limits that represent 100% containment limits.

H.2.2 Outlet Gas Pressure ($\varepsilon_{P_{out,b}}$ and $\varepsilon_{P_{out,r}}$)

The Wallace & Tiernan FA-139 Precision Aneroid Barometer has a span of 13.75 to 15.25 psia (28 to 30 in Hg). The accuracy specification of the barometer is $\pm 0.3\%$ of full scale.[98] In this analysis, the manufacturer specified accuracy is interpreted to be

$$\pm \frac{0.3}{100} \times (15.25 - 13.75)\,\text{psia} = \pm 0.003 \times 1.5\,\text{psia} = \pm 0.0045\,\text{psia}.$$

---

[97] Wyle Reference Standards Laboratory Report of Test Number 6.85331.02, June 24, 2008.

[98] Wallace & Tiernan Technical Data Sheet – Precision Aneroid Barometer, Types FA-112, FA-139, FA-160, FA-185, Cat. File 610.100, Revised 7-89.

The bias in the barometer pressure is assumed to follow a normal distribution and the accuracy limits are assumed to represent 95% containment limits.

The FA-139 barometer has an analog resolution of 0.005 psi (0.01 in Hg). The resolution error limits are ± 0.0025 psi and are assumed to represent 95% containment (confidence) limits. The analog resolution error is assumed to follow a normal distribution.

H.2.3 Outlet Gas Temperature ($\varepsilon_{T_{out,b}}$ and $\varepsilon_{T_{out,r}}$)

The Rosemount model 162C PRT has an accuracy of ± 0.22 °C and it's output is read with an Instrulab RTD monitor that has a digital resolution of 0.01 °C.[99] The bias in the PRT temperature is assumed to follow a normal distribution. The accuracy limits are assumed to correspond to 95% containment limits.

The digital resolution error limits are ± 0.005 °C. The digital resolution error follows a uniform distribution with an associated 100% containment probability.

H.2.4 Uncertainty in R_{corr}

The uncertainty in the corrected rate is equal to the square root of the distribution variance for $\varepsilon_{R_{corr}}$.

$$u_{\varepsilon_{R_{corr}}} = \sqrt{\mathrm{var}(\varepsilon_{R_{corr}})} \tag{H-11}$$

Applying the variance operator to equation (H-10), and noting that there are no correlations between error sources, gives

$$
\begin{aligned}
u_{\varepsilon_{R_{corr}}} &= \sqrt{\mathrm{var}\left(\varepsilon_{R_{corr}}\right)} \\
&= \sqrt{\begin{array}{l} c_{R_{ref}}^2 \, \mathrm{var}\left(\varepsilon_{R_{ref},b}\right) + c_{R_{ref}}^2 \, \mathrm{var}\left(\varepsilon_{R_{ref},r}\right) + c_{P_{out}}^2 \, \mathrm{var}\left(\varepsilon_{P_{out,b}}\right) \\ + c_{P_{out}}^2 \, \mathrm{var}\left(\varepsilon_{P_{out,r}}\right) + c_{T_{out}}^2 \, \mathrm{var}\left(\varepsilon_{T_{out,b}}\right) + c_{T_{out}}^2 \, \mathrm{var}\left(\varepsilon_{T_{out,r}}\right) \end{array}}
\end{aligned} \tag{H-12}
$$

The variance terms in equation (H-12) are equivalent to the square of the uncertainty in the corresponding error (e.g., $\mathrm{var}(\varepsilon_{R_{ref},b}) = u_{\varepsilon_{R_{ref},b}}^2$). So, equation (h-12) can be rewritten in terms of the individual measurement process uncertainties.

$$
u_{\varepsilon_{R_{corr}}} = \sqrt{\begin{array}{l} c_{R_{ref}}^2 \, u_{\varepsilon_{R_{ref},b}}^2 + c_{R_{ref}}^2 \, u_{\varepsilon_{R_{ref},r}}^2 + c_{P_{out}}^2 \, u_{\varepsilon_{P_{out,b}}}^2 + c_{P_{out}}^2 \, u_{\varepsilon_{P_{out,r}}}^2 \\ + c_{T_{out}}^2 \, u_{\varepsilon_{T_{out,b}}}^2 + c_{T_{out}}^2 \, u_{\varepsilon_{T_{out,r}}}^2 \end{array}} \tag{H-13}
$$

The partial derivative equations used to compute the sensitivity coefficients are given in

---

[99] Wyle Reference Standards Laboratory Report of Test Number 6.85331.02, June 24, 2008.

equations (H-14) through (H-16).

$$c_{R_{ref}} = \frac{\partial R_{corr}}{\partial R_{ref}} = \left(\frac{P_{std}}{P_{out}} \times \frac{T_{out}}{T_{std}} \right)^{1/2} \tag{H-14}$$

$$c_{P_{out}} = \frac{\partial R_{corr}}{\partial P_{out}} = -\frac{1}{2} R_{ref} \times \left(\frac{P_{std} \times T_{out}}{T_{std}} \right)^{1/2} \times P_{out}^{-3/2} \tag{H-15}$$

$$c_{T_{out}} = \frac{\partial R_{corr}}{\partial T_{out}} = \frac{1}{2} R_{ref} \times \left(\frac{P_{std}}{P_{out} \times T_{std}} \right)^{1/2} \times T_{out}^{-1/2} \tag{H-16}$$

The measurement process uncertainties are estimated from the specification limits, containment probability (confidence level) and the inverse error distribution function.

The measurement reference bias uncertainty is estimated using the ± 0.5% or reading tolerance limits, the inverse normal distribution function, Φ^{-1}, and a 0.95 containment probability (95% confidence level). The bias uncertainty in a measurement reference flow rate of 2937.12 sccm is computed for illustrative purposes.

$$u_{\varepsilon_{R_{ref},b}} = \frac{0.005 \times 2937.12 \, \text{sccm}}{\Phi^{-1}\left(\frac{1+0.95}{2} \right)} = \frac{14.69 \, \text{sccm}}{1.9600} = 7.49 \, \text{sccm}$$

The measurement reference digital resolution uncertainty is estimated using the ± 0.0005 sccm tolerance limits, the inverse uniform distribution function and a 1.00 containment probability (100% confidence level).

$$u_{\varepsilon_{R_{ref},r}} = \frac{0.0005 \, \text{sccm}}{\sqrt{3}} = \frac{0.0005 \, \text{sccm}}{1.732} = 2.89 \times 10^{-4} \, \text{sccm}$$

The barometer bias uncertainty is estimated using the ± 0.0045 psia tolerance limits, the inverse normal distribution function, Φ^{-1}, and a 0.95 containment probability (95% confidence level).

$$u_{\varepsilon_{P_{out},b}} = \frac{0.0045 \, \text{psia}}{\Phi^{-1}\left(\frac{1+0.95}{2} \right)} = \frac{0.0045 \, \text{psia}}{1.9600} = 0.0023 \, \text{psia}$$

The barometer resolution uncertainty is estimated using the ± 0.0025 psia tolerance limits, the inverse normal distribution function, Φ^{-1}, and a 0.95 containment probability (95% confidence level).

228

$$u_{\varepsilon_{Pout,r}} = \frac{0.0025\,\text{psia}}{\Phi^{-1}\left(\dfrac{1+0.95}{2}\right)} = \frac{0.0025\,\text{psia}}{1.9600} = 0.0013\,\text{psia}$$

The PRT bias uncertainty is estimated using the ± 0.22 °C tolerance limits, the inverse normal distribution function, Φ^{-1}, and a 0.95 containment probability (95% confidence level).

$$u_{\varepsilon_{Tout,b}} = \frac{0.22\,^{\circ}\text{C}}{\Phi^{-1}\left(\dfrac{1+0.95}{2}\right)} = \frac{0.22\,^{\circ}\text{C}}{1.9600} = 0.11\,^{\circ}\text{C}$$

The PRT digital resolution uncertainty is estimated using the ± 0.005 °C digital resolution limits, the inverse uniform distribution function and a 1.00 containment probability (100% confidence level).

$$u_{\varepsilon_{Tout,r}} = \frac{0.005\,^{\circ}\text{C}}{\sqrt{3}} = \frac{0.005\,^{\circ}\text{C}}{1.732} = 0.003\,^{\circ}\text{C}$$

The estimated measurement process uncertainties and sensitivity coefficients are summarized in Table H-2. The component uncertainty is the product of the standard uncertainty and the sensitivity coefficient.

Table H-2. Measurement Process Uncertainties for Corrected Flow Rate = 2931.21 sccm

Error Source	± Error Limits	Error Distribution	Confid. Level	Standard Uncertainty	Sensitivity Coefficient	Component Uncertainty
$u_{\varepsilon_{Rref,b}}$	± 14.69 sccm	Normal	95	7.49 sccm	0.9980	7.47 sccm
$u_{\varepsilon_{Rref,r}}$	± 0.0005 sccm	Uniform	100	2.89×10^{-4} sccm	0.9980	2.88×10^{-4} sccm
$u_{\varepsilon_{Pout,b}}$	± 0.0045 psia	Normal	95	0.0023 psia	- 98.71 sccm/psia	0.227 sccm
$u_{\varepsilon_{Pout,r}}$	± 0.0025 psia	Normal	95	0.0013 psia	- 98.71 sccm/psia	0.128 sccm
$u_{\varepsilon_{Tout,b}}$	± 0.22 °C	Normal	95	0.11 °C	4.95 sccm/°C	0.545 sccm
$u_{\varepsilon_{Tout,r}}$	± 0.005 °C	Uniform	100	0.003 °C	4.95 sccm/°C	0.015 sccm

The uncertainty in R_{corr} is computed by taking the root sum square of the component uncertainties.

$$
\begin{aligned}
u_{\varepsilon_{Rcorr}} &= \sqrt{(7.47)^2 + \left(2.88\times10^{-4}\right)^2 + (0.227)^2 + (0.128)^2 + (0.545)^2 + (0.015)^2}\;\text{sccm}\\
&= \sqrt{55.801 + 8.29\times10^{-8} + 0.0515 + 0.0164 + 0.2970 + 2.25\times10^{-4}}\;\text{sccm}\\
&= \sqrt{56.166}\;\text{sccm} = 7.49\;\text{sccm}
\end{aligned}
$$

The Welch-Satterthwaite formula is used to compute the degrees of freedom for $u_{\varepsilon_{R_{corr}}}$.

$$\nu_{u_{\varepsilon_{R_{corr}}}} = \cfrac{u^4_{\varepsilon_{R_{corr}}}}{\left[\begin{array}{l} \cfrac{c^4_{R_{ref}} u^4_{\varepsilon_{R_{ref},b}}}{\nu_{u_{\varepsilon_{R_{ref},b}}}} + \cfrac{c^4_{R_{ref}} u^4_{\varepsilon_{R_{ref},r}}}{\nu_{u_{\varepsilon_{R_{ref},r}}}} + \cfrac{c^4_{P_{out}} u^4_{\varepsilon_{P_{out},b}}}{\nu_{u_{\varepsilon_{P_{out},b}}}} \\[2em] + \cfrac{c^4_{P_{out}} u^4_{\varepsilon_{P_{out},r}}}{\nu_{u_{\varepsilon_{P_{out},r}}}} + \cfrac{c^4_{T_{out}} u^4_{\varepsilon_{T_{out},b}}}{\nu_{u_{\varepsilon_{T_{out},b}}}} + \cfrac{c^4_{T_{out}} u^4_{\varepsilon_{T_{out},r}}}{\nu_{u_{\varepsilon_{T_{out},r}}}} \end{array}\right]} \qquad \text{(H-17)}$$

The degrees of freedom for all of the process uncertainties are infinite. Therefore, the degrees of freedom for $u_{\varepsilon_{R_{corr}}}$ are also infinite.

The pareto chart, shown in Figure H-2, indicates that the measurement reference bias uncertainty is the largest contributor to the uncertainty in R_{corr}.

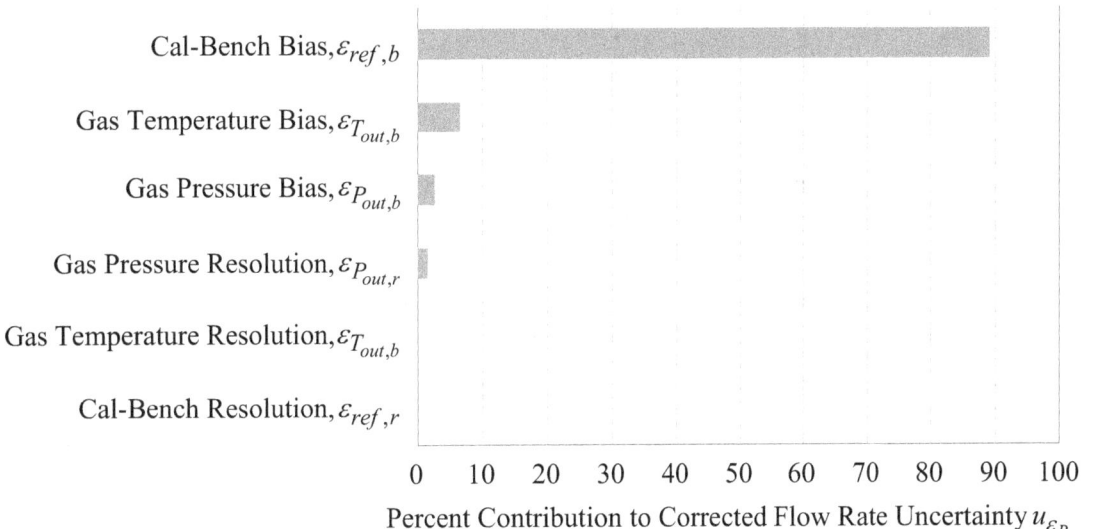

Figure H-2. Pareto Chart for Corrected Flow Rate Uncertainty

Uncertainty estimates for all of the corrected flow rate data are summarized in Table H-3.

Table H-3. Uncertainty Estimates for Corrected Measurement Reference Flow Rates

UUT Rotameter Scale Reading	Outlet Gas Pressure (psia)	Outlet Gas Temperature (°C)	Measurement Reference Flow Rate (sccm)	Corrected Flow Rate R_{corr} (ccm)	Standard Uncertainty $u_{\varepsilon_{R_{corr}}}$ (ccm)	Degrees of Freedom
15.00	14.850	23.1	548.69	547.68	1.40	∞
30.00	14.850	23.1	1356.75	1354.25	3.47	∞
60.00	14.850	23.0	2937.12	2931.21	7.49	∞
90.00	14.855	22.9	4551.52	4540.84	11.62	∞

UUT Rotameter Scale Reading	Outlet Gas Pressure (psia)	Outlet Gas Temperature (°C)	Measurement Reference Flow Rate (sccm)	Corrected Flow Rate R_{corr} (ccm)	Standard Uncertainty $u_{\varepsilon R_{corr}}$ (ccm)	Degrees of Freedom
120.00	14.855	22.9	6227.96	6213.34	15.90	∞
150.00	14.860	22.6	7881.43	7857.62	20.11	∞

H.3 Predicted Flow Rate Uncertainty

As previously discussed, the rotameter is calibrated to establish a third-order polynomial equation that can be used to predict gas flow rates. Ideally, a weighted least squares regression analysis would be conducted to establish the appropriate equation coefficients. The solution for the coefficients are estimated using the following matrix equation

$$\mathbf{b} = (\mathbf{X'WX})^{-1}\,\mathbf{X'WY} \tag{H-18}$$

where

$$\mathbf{b} = \begin{pmatrix} b_0 \\ b_1 \\ b_2 \\ b_3 \end{pmatrix} \qquad \mathbf{X} = \begin{pmatrix} 1 & x_1 & x_1^2 & x_1^3 \\ 1 & x_2 & x_2^2 & x_2^3 \\ M & M & M & M \\ 1 & x_k & x_k^2 & x_k^3 \end{pmatrix} \qquad \mathbf{W} = \begin{pmatrix} w_1 & & & 0 \\ & w_2 & & \\ & & O & \\ 0 & & & w_k \end{pmatrix} \qquad \mathbf{Y} = \begin{pmatrix} y_1 \\ y_2 \\ M \\ y_k \end{pmatrix}$$

k is the number of x data points and $\mathbf{X'}$ is the transpose of $\mathbf{X}$. The weighting factor matrix $\mathbf{W}$ consists of the estimated uncertainties for the y measured values.

In the absence of a dedicated regression analysis software program or a sophisticated statistical analysis package, a non-weighted regression fit is often obtained using a standard spreadsheet application.

The non-weighted third-order polynomial equation obtained by applying the Microsoft Excel **Add Trendline** function for the corrected flow rate versus rotameter scale reading data is given in equation (H-19).

$$R_{calc} = -218.667 + 51.325x + 0.022x^2 - 0.000034x^3 \tag{H-19}$$

where

R_{calc} = calculated or predicted flow rate in cubic centimeter per minute (ccm)
x = rotameter scale reading

The error equation for R_{calc} is the sum of the corrected flow rate error and the regression fit error.

$$\varepsilon_{R_{calc}} = \varepsilon_{R_{corr}} + \varepsilon_{reg} \tag{H-20}$$

The uncertainty in the calculated rate is equal to the square root of the distribution variance for $\varepsilon_{R_{calc}}$.

$$u_{\varepsilon_{R_{calc}}} = \sqrt{\text{var}(\varepsilon_{R_{calc}})} = \sqrt{\text{var}\left(\varepsilon_{R_{corr}} + \varepsilon_{reg}\right)} \qquad \text{(H-21)}$$

Noting that there are no correlations between the corrected flow rate and regression fit errors, the uncertainty in the calculated or predicted flow rate is computed from equation (H-22).

$$u_{\varepsilon_{R_{calc}}} = \sqrt{\text{var}\left(\varepsilon_{R_{corr}}\right) + \text{var}\left(\varepsilon_{reg}\right)} = \sqrt{u^2_{\varepsilon_{R_{corr}}} + u^2_{\varepsilon_{reg}}} \qquad \text{(H-22)}$$

The uncertainty due the regression fit is called the *standard error of forecast*.[100] The standard error of forecast accounts for the fact that regression equation (H-19) was generated from a finite sample of data. If another sample of data were collected, then a different regression equation would result. The standard error or forecast considers the dispersion of various regression equations that would be generated from multiple sample sets around the true population regression equation. The standard error of forecast is computed from

$$s_f = s_{y,x}\sqrt{1 + \mathbf{x}'\left(\mathbf{X}'\mathbf{W}\mathbf{X}\right)^{-1}\mathbf{x}} \qquad \text{(H-23)}$$

where $s_{y,x}$ is the standard error of estimate, $\mathbf{x}'$ is the transpose of $\mathbf{x}$ and

$$\mathbf{X} = \begin{pmatrix} 1 \\ x \\ x^2 \\ x^3 \end{pmatrix} \qquad \text{(H-24)}$$

The standard error of estimate is a measure of the difference between actual values and values estimated from a given regression equation. The standard error of estimate is also defined as the standard deviation of the normal distributions of y for any given x. The standard error of estimate is computed from

$$s_{y,x} = \sqrt{\frac{\sum(y - \hat{y})^2}{k - (m+1)}} \qquad \text{(H-25)}$$

where $\hat{y}$ is the predicted or calculated value and m is the order of the regression equation (i.e., m = 1, 2, 3 or higher). A regression analysis that has a small standard error of estimate has data points that are very close to the regression line. Conversely, a large standard error of estimate results when data points are widely dispersed around the regression line. The degrees of freedom for $s_{y,x}$ is

$$\nu_{s_{y,x}} = k - (m+1) \qquad \text{(H-26)}$$

The standard error of estimate for the flow rate equation (H-19) is computed to be

---

[100] Hanke, J. et al: *Statistical Decision Models for Management*, Allyn and Bacon, Inc. 1984.

$$s_{y,x} = \sqrt{\frac{\sum(y-\hat{y})^2}{k-(3+1)}} = \sqrt{\frac{676.73}{6-4}} \text{ ccm} = \sqrt{338.37} \text{ ccm} = 18.4 \text{ ccm}$$

and the associated degrees of freedom are

$$\nu_{s_{y,x}} = k - (m+1) = 6 - (3+1) = 2$$

The degrees of freedom for s_f are equal to those for $s_{y,x}$.

The data used to compute the standard error of estimate are listed in Table H-4. The standard error of forecast for each value of x is also listed.

Table H-4. Standard Error of Forecast for Regression Equation (H-19)

Rotameter Scale Reading x	Corrected Flow Rate y (ccm)	Predicted or Calculated Flow Rate $\hat{y}$ (ccm)	$y - \hat{y}$ (ccm)	$(y-\hat{y})^2$ (ccm²)	Standard Error of Forecast s_f (ccm)	Degrees of Freedom ν_{s_f}
15.00	547.68	556.06	- 8.38	70.22	25.0	2
30.00	1354.25	1340.05	14.2	201.64	22.1	2
60.00	2931.21	2933.01	-1.80	3.24	23.3	2
90.00	4540.84	4554.65	- 13.81	190.72	22.2	3
120.00	6213.34	6199.42	13.92	193.77	23.7	4
150.00	7857.62	7861.76	- 4.14	17.14	25.8	5
$\bar{x} = 77.50$			$\sum(y-\hat{y})^2 = 676.73$			

Setting $u_{\varepsilon_{reg}}$ equal to s_f, the uncertainties in the calculated rates can now be computed using equation (H-22). For example, the uncertainty in the calculated rate $R_{cal} = 2933.01$ ccm is

$$u_{\varepsilon_{R_{calc}}} = \sqrt{u^2_{\varepsilon_{R_{corr}}} + u^2_{\varepsilon_{reg}}}$$

$$= \sqrt{(7.49)^2 + (23.3)^2} \text{ ccm}$$

$$= \sqrt{598.99} \text{ ccm} = 24.5 \text{ ccm}$$

The degrees of freedom for $u_{\varepsilon_{R_{calc}}}$ are computed using the Welch-Satterthwaite formula.

$$v_{u_{\varepsilon_{Rcalc}}} = \frac{u_{\varepsilon_{Rcalc}}^4}{\dfrac{u_{\varepsilon_{Rcorr}}^4}{v_{u_{\varepsilon_{Rcorr}}}} + \dfrac{u_{\varepsilon_{reg}}^4}{v_{u_{\varepsilon_{reg}}}}} = \frac{u_{\varepsilon_{Rcalc}}^4}{\dfrac{u_{\varepsilon_{Rcorr}}^4}{\infty} + \dfrac{u_{\varepsilon_{reg}}^4}{2}} = 2 \times \frac{u_{\varepsilon_{Rcalc}}^4}{u_{\varepsilon_{reg}}^4} \tag{H-27}$$

The degrees of freedom are expressed as the nearest whole number value. For example, the degrees of freedom for $u_{\varepsilon_{Rcalc}} = 24.5\,ccm$ are computed to be

$$v_{u_{\varepsilon_{Rcalc}}} = 2 \times \frac{u_{\varepsilon_{Rcalc}}^4}{u_{\varepsilon_{reg}}^4} = 2 \times \left(\frac{24.5}{23.3}\right)^4 = 2 \times (1.05)^4$$

$$= 2 \times 1.22 = 2.44 = 2$$

The confidence limits for R_{calc} can be expressed as

$$R_{calc} \pm t_{\alpha/2,v} \times u_{\varepsilon_{Rcalc}} \tag{H-28}$$

where $t_{\alpha/2,v}$ is the Student's t-statistic. For a 95% confidence level, $t_{0.0025,2} = 4.3027$ and the confidence limits for $R_{cal} = 2933.01$ ccm are computed to be

$$2933.01\,ccm \pm 4.3027 \times 24.5\,ccm$$
$$\text{or}$$
$$2933.01\,ccm \pm 105.31\,ccm$$

The above confidence limits can also be expressed as a percentage of the full scale (FS) output of the rotameter.

$$2933.01\,ccm \pm \frac{105.31\,ccm}{7861.76\,ccm} \times 100\%$$
$$\text{or}$$
$$2933.01\,ccm \pm 1.34\%$$

The computed uncertainties, degrees of freedom and 95% confidence limits for calculated rates at six rotameter readings are listed in Table H-5. The manufacturer specified accuracy of the Series 1110 rotameter is ± 2% FS.[101] The 95% confidence limits computed for the UUT rotameter fall within the accuracy specifications.

---

[101] Design Specifications DS-1110-1140 for 1110 and 1140 Series Glass Tube Full-View Flowmeters, Brooks Instruments, January 1998.

Table H-5. Uncertainties for Calculated Flow Rates

Rotameter Scale Reading x	Calculated Flow Rate R_{calc} (ccm)	Corrected Flow Rate Uncert. $u_{\varepsilon_{R_{corr}}}$ (ccm)	Regression Uncert. $u_{\varepsilon_{reg}}$ (ccm)	Calculated Flow Rate Uncert. $u_{\varepsilon_{R_{calc}}}$ (ccm)	Degrees of Freedom $\nu_{u_{\varepsilon_{R_{calc}}}}$	Student's t-statistic $t_{\alpha/2,\nu}$	95% Conf. Limits (ccm)	95% Conf. Limits (% FS)
15.00	556.06	1.40	25.0	25.0	2	4.3027	± 107.74	± 1.37
30.00	1340.05	3.47	22.1	22.4	2	4.3027	± 96.25	± 1.22
60.00	2933.01	7.49	23.3	24.5	2	4.3027	± 105.31	± 1.34
90.00	4554.65	11.62	22.2	25.1	3	3.1824	± 79.74	± 1.01
120.00	6199.42	15.90	23.7	28.5	4	2.7765	± 79.24	± 1.01
150.00	7861.76	20.11	25.8	32.7	5	2.5706	± 84.09	± 1.07

APPENDIX I – WINGBOOM AOA ANALYSIS EXAMPLE

The purpose of this analysis is to estimate the overall uncertainty in an aircraft wingboom angle of attack (AOA) measurement. The wingboom AOA measurement uncertainty is estimated using the system analysis procedure discussed in Chapter 7.

I.1   Measurement Process Overview

A BEI Model 1201 5k Ohm potentiometer, with a maximum rotational travel of 354°, is the primary sensor used to measure the wingboom AOA. The potentiometer output voltage is run through a SCD-108S signal conditioning card manufactured by Teletronics Technology Corporation (TTC). The signal conditioning card consists of an 8-channel multiplexer, amplifier, low-pass filter, and analog to digital converter (ADC).

The ADC uses 12-bit precision to convert the continuous voltage signal to a binary code. Therefore, the output signal from the ADC is a quantized value ranging from 0 to 4095 counts (i.e., $2^{12} - 1$).[102] The ADC counts output is converted back to a wingboom angle using a linear equation obtained from a regression fit of calibration data.

The wingboom AOA measurement system is calibrated from - 45° to + 45° using an E-2C 535 Boom Universal Calibrator Fixture. The calibrator fixture is, in turn, calibrated according to the LIST-A020 procedure.[103] The wingboom AOA calibration data[104] are listed in Table I-1.

Table I-1. Wingboom AOA Calibration Data

Meas. Number	ADC Counts	Applied Angle	Meas. Number	ADC Counts	Applied Angle
1	2052	0.875	18	2714	15.875
2	1878	-4.125	19	2931	20.875
3	1630	-9.125	20	3144	25.875
4	1378	-14.125	21	3361	30.875
5	1142	-19.125	22	3575	35.875
6	716	-29.125	23	3789	40.875
7	289	-39.125	24	4002	45.875
8	76	-44.125	25	3788	40.875
9	289	-39.125	26	3575	35.875
10	717	-29.125	27	3361	30.875
11	1142	-19.125	28	3144	25.875
12	1378	-14.125	29	2929	20.875
13	1629	-9.125	30	2712	15.875
14	1878	-4.125	31	2494	10.875
15	2052	0.875	32	2274	5.875
16	2273	5.875	33	2053	0.875
17	2496	10.875			

---

[102] Email from Kenneth Miller, CIV NAVAIR to Dr. Howard Castrup, Integrated Sciences Group, Sent: 7/13/04 Subject: LSBF Coefficient Significant Digits.

[103] Naval Air Test Center Technical Manual, Local Calibration Procedure LIST-A020, 1 November 2003.

[104] Calibration Data Sheet, C-2A, 162142 NP2000, TMATS File: H:\projects\C2 NP2000\C2np2k07.tma

A regression analysis was conducted to obtain an unweighted least squares best fit (LSBF) to a straight line, as shown in Figure I-1.

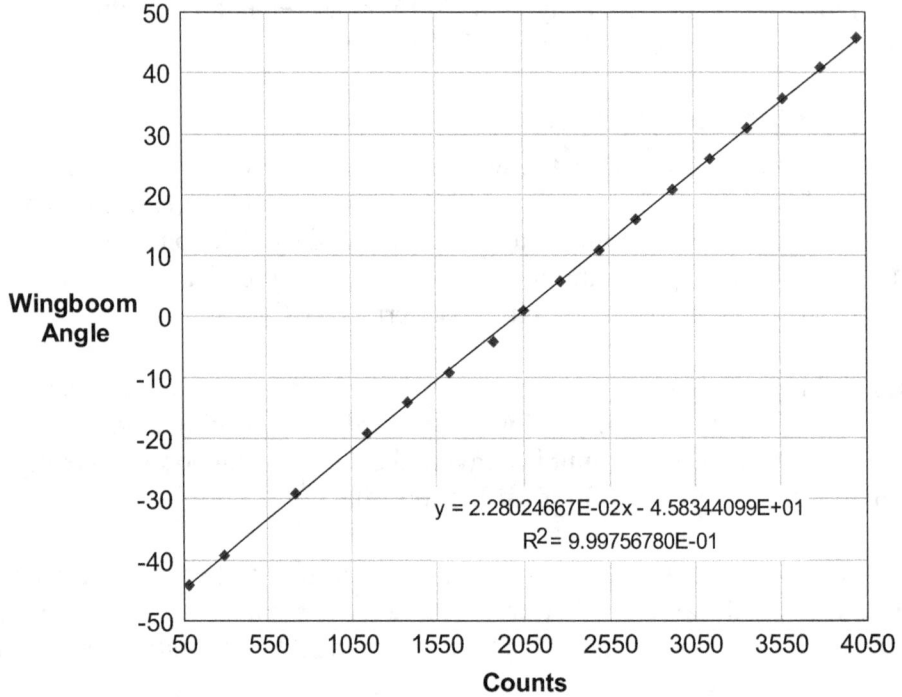

$$y = 2.28024667E\text{-}02x - 4.58344099E\text{+}01$$
$$R^2 = 9.99756780E\text{-}01$$

Figure I-1. Straight Line Fit of Calibration Data

The straight line fit equation (I-1) is used to convert the recorded counts data to wingboom angle.

$$\text{Wingboom Angle} = 0.0228 \times \text{Counts} - 45.83 \tag{I-1}$$

I.2   System Model

The wingboom AOA measurement is made through a linear sequences of stages or modules as shown in Figure I-2. The output, *Y*, from any given system module comprises the input of the next module in the series. Since each module's output carries with it an element of uncertainty, this means that this uncertainty will be present at the input of a subsequent module.

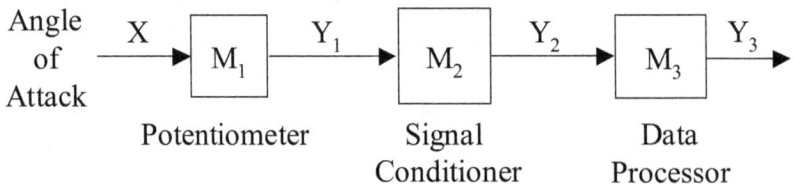

Figure I-2. Block Diagram of Wingboom AOA Measurement System

I.3   System Input

In this example, a nominal wingboom AOA of 20° will be analyzed. The calibrator fixture is used to provide the wingboom AOA. Therefore, any uncertainty in the angle established by the calibrator fixture must be determined and included in this analysis. The following error sources are considered relevant for the calibration fixture angle:

- Bias of calibrator fixture angle
- Measurement repeatability or random error

I.3.1 Calibrator Fixture Bias

The E-2C 535 calibrator fixture is reported to have tolerance limits of ± 0.25° of the angle established via the LIST-A020 procedure.[105] For the purposes of this analysis, these limits are assumed to represent a 99% confidence limits for a normally distributed error.

I.3.2 Measurement Repeatability

Repeatability or random error results from variations that are manifested through repeat wingboom AOA measurements over a short time period. Repeatability uncertainty can have units of the potentiometer output or signal conditioner output depending on the calibration procedure used. As seen from the calibration data listed in Table I-1, the LIST-A020 procedure calibrates the potentiometer and signal conditioner as a combined unit. Therefore, repeatability should be evaluated as an error source in the signal conditioning module (M_2).

I.4 System Modules

The following subsections describe the measurement system modules in detail, identifying error sources and defining appropriate module output equations. Manufacturer specifications will be used to establish error limits. Manufacturer specification documents, as well as other reference materials used in this analysis, are listed in the footnotes.

I.4.1 Potentiometer Module (M_1)

The first module consists of the Model 1201 5k Ohm potentiometer manufactured by BEI Technologies, Inc. Potentiometers are essentially a resistor, R_P, connected to a voltage source, V_I, with a moving contact or wiper.[106] The resistor is "divided" at the point of wiper contact and the voltage output signal, V_O, is proportional to the voltage drop across the resulting load resistance, R_L, as shown in Figure I-3a.

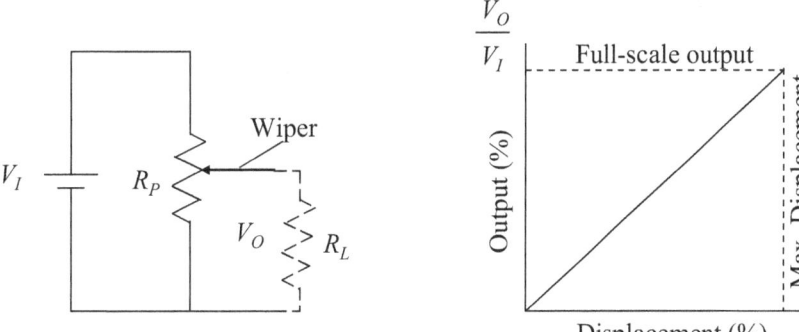

Figure I-3. a. Potentiometer Circuit   b. Ideal Linear Response Characteristic

---

[105] Naval Air Test Center Technical Manual, Local Calibration Procedure LIST-A020, 1 November 2003.

[106] Measurement, Instrumentation, and Sensors Handbook, CRCnetBase 1999, John G. Webster Editor-in Chief.

Potentiometers are commonly designed to generate a DC voltage output that is linearly proportional to rotational or lateral displacement X/X_p, as shown in Figure I-3b. The potentiometer voltage output is then expressed as

$$V_o = V_I \times K \times \frac{X}{X_P} \qquad (\text{I-2})$$

where K is the potentiometer sensitivity, X is the input angle in degrees and X_p is the maximum angle that the potentiometer can travel.

When developing an uncertainty analysis for the potentiometer module, the impact that errors in V_I, K, X and X_p will have on the output value must considered. In addition, manufacturer specifications for the Model 1201 5k Ohm potentiometer[107] indicate that there are other error sources that affect the potentiometer output. The following error sources are applicable to the output of the potentiometer module:

- Calibrator Fixture Angle
- Supply Voltage
- Maximum Angle
- Sensitivity
- Linearity
- Resistance
- Noise
- Resolution
- Temperature Coefficient

I.4.1.1   Calibrator Fixture Angle (ε_X)

As previously discussed, a nominal angle of 20 is applied by the calibrator fixture. The calibrator fixture is reported to have tolerance limits of $\pm$ 0.25° of the applied angle. These limits are assumed to represent a 99% confidence limits for a normally distributed error.

I.4.1.2   Supply Voltage (ε_{V_I})

Since the potentiometer is a passive sensor, the signal conditioner must provide a regulated DC voltage or current via a precision power supply. The SCD-108S signal conditioner supplies an external excitation voltage of 5 V to the potentiometer. The excitation voltage accuracy is stated to be $\pm 0.3\%$ maximum. In this analysis these limits are assumed to represent 95% confidence limits for a normally distributed error.

I.4.1.3   Maximum Angle (ε_{X_P})

The specification sheet for the BEI Model 1201 potentiometer indicates a maximum or actual electrical travel of 354° $\pm$ 2°. We interpret the $\pm$ 2° limits to represent 95% confidence limits for a normally distributed error.

---

[107] Specification sheet for BEI Model 1201 Servo Mount Wirewound Single-turn Precision Potentiometer, www.beiduncan.com.

I.4.1.4   Sensitivity (ε_K)

The potentiometer sensitivity is the dimensionless slope of the linear response curve shown in Figure I-3b. Ideally, the potentiometer sensitivity should have a value of unity. However, variation in potentiometer sensitivity can occur due to temperature effects, drift, hysteresis or other factors.

I.4.1.4.1 Linearity (ε_L)

Linearity, or more appropriately non-linearity, is a measure of the deviation of the actual input-to-output performance of the device from an ideal linear relationship. Linearity error is fixed at any given input, but varies with magnitude and sign over a range of inputs. Therefore, it is considered to be a normally distributed error.

The specification sheet for the BEI Model 1201 indicates linearity tolerance limits of $\pm$ 0.5% of full scale (FS) for standard conditions and $\pm$ 0.2% FS for best practical conditions. The $\pm$ 0.5% FS limits are used in this analysis and are be assumed to represent 95% confidence limits for a normally distributed error.

I.4.1.4.2 Resistance (ε_R)

Total resistance is a key parameter because it determines the amount of current drawn for a given applied voltage. Because potentiometer resistance can change over time between calibrations, it is important to estimate how resistance error impacts overall uncertainty.

Manufacturer specifications indicate that the resistance tolerance limits for the BEI Model 1201 are $\pm$ 3% FS for standard conditions and $\pm$ 1% FS for best practical conditions. The $\pm$ 3% FS limits are used in this analysis and are assumed to represent 95% confidence limits for a normally distributed error.

I.4.1.4.3 Noise (ε_N)

Non-repeatability or random error intrinsic to the device, that causes the output to vary from observation to observation for a constant input, is usually specified as noise.

Manufacturer specifications indicate that the equivalent noise resistance (ENR) has a maximum value of 100 Ohms. Potentiometer noise, in relation to the total potentiometer resistance of 5,000 Ohms, can be expressed as $\pm$ 2% of FS. In this analysis, the $\pm$ 2% of FS limits are assumed to represent 95% confidence limits for a normally distributed error.

I.4.1.4.4 Resolution (ε_{res})

Resolution defines the smallest possible increment of voltage change that can be produced and detected. In wire-wound coil potentiometers, resolution is the voltage drop in one turn of resistance wire. The best attainable resolution is $1/N \times 100\%$ of full scale voltage or resistance, where N is the number of turns in the coil. Resolution can also be expressed in terms of travel in inches or degrees.

The specification sheet for the BEI Model 1201 contains a footnote that indicates that Resolution Tables are available by model number and resistance value. The manufacturer was contacted and they indicated resolution limits of ± 0.11% FS. These limits are assumed to represent 100% confidence limits for a uniformly distributed error.

I.4.1.4.5 Temperature Coefficient (ε_{TC})

Resistance increases with temperature. Therefore, the potentiometer sensitivity will be affected by temperature variation. However, this may not be a major concern as long as the changes in resistance are uniform and the potentiometer is operated within its rated temperature range. In general, wire-wound potentiometers have very low temperature coefficients.

The temperature coefficient tolerance limits specified for the BEI Model 1201 are ± 0.007%/°C. These limits are assumed to represent 95% confidence limits for a normally distributed error.

A temperature range of 50°C with associated error limits of ± 2 °C are used in this analysis. The ± 2 °C limits are assumed to represent 95% confidence limits for a normally distributed error.

I.4.1.4.6 Potentiometer Output Equation

The output equation for the potentiometer module is expressed in equation (I-3). Table I-2 contains the relevant information for the equation parameters.

$$P_{out} + \varepsilon_{P_{out}} = V_O + \varepsilon_{V_O} = \left(V_I + \varepsilon_{V_I}\right)\left(K + \varepsilon_K\right)\frac{X + \varepsilon_X}{X_P + \varepsilon_{X_P}} \tag{I-3}$$

where

$$\begin{aligned} \varepsilon_K &= \varepsilon_L + \varepsilon_R + \varepsilon_N + \varepsilon_{res} + \varepsilon_{TC} \times \left(\Delta T + \varepsilon_{\Delta T}\right) \\ &= \varepsilon_L + \varepsilon_R + \varepsilon_N + \varepsilon_{res} + \varepsilon_{TC} \times \Delta T + \varepsilon_{TC} \times \varepsilon_{\Delta T} \end{aligned} \tag{I-4}$$

and ΔT and $\varepsilon_{\Delta T}$ are the temperature range and corresponding error, respectively.

Table I-2. Parameters used in Potentiometer Module Equation

Equation Parameter	Description	Nominal or Mean Value	Error Limits	Percent Confidence	Error Distribution
V_I	Supply Voltage	5 V			
ε_{V_I}	Supply Voltage Error	0 V	± 15 mV	95	Normal
X	Calibrator Fixture Angle	20°			
ε_X	Fixture Angle Error	0°	± 0.25°	99	Normal
X_P	Maximum Angle	354°			
ε_{X_P}	Max. Angle Error	0°	± 2°	95	Normal
K	Potentiometer Sensitivity	1.0			
ε_L	Sensitivity Linearity	0	± 0.005	95	Normal
ε_R	Resistance Error	0	± 0.03	95	Normal
ε_N	Noise	0	± 0.02	95	Normal
ε_{res}	Resolution Error	0	± 0.0011	100	Uniform
ε_{TC}	Temperature Coefficient	0	± 0.7e^{-4}/°C	95	Normal
ΔT	Temperature Range	50 °C			

$\varepsilon_{\Delta T}$	Temperature error	0 °C	± 2 °C	95	Normal

I.4.2   Signal Conditioner Module (M₂)

The voltage signal entering the SCD-108S signal conditioning card is converted to a quantized value ranging from 0 to 4095 counts (i.e., $2^{12}-1$). The manufacturer specifications for the SCD-108S signal conditioning card[108] state an accuracy of $\pm 0.5\%$. As previously discussed in Section I.3.2, repeatability or random error must also be considered in the analysis of the signal conditioner module.

The following error sources are applicable to the output of the signal conditioner module:

- Signal conditioner bias
- Quantization error
- Measurement repeatability

I.4.2.1 Signal Conditioner Bias (ε_{SC_b})

TTC, the manufacturer of the SCD-108S, was contacted to obtain clarification regarding the accuracy specification limits. TTC stated the accuracy limits are a percent full scale output and that the associated confidence level is 99%. The full scale output of the SCD-108S is 4095 counts. Therefore, the accuracy limits are ± 20.475 counts. In this analysis, the signal conditioner bias is assumed to be normally distributed.

I.4.2.2 Quantization Error (ε_{SC_q})

During quantization, a finite number is used to represent a continuous value. The resulting resolution limit from the quantization of a 5 V signal using a 12-bit ADC is 5 V/(2^{12}) or 1.2 mV. The quantization error limits are half the resolution or ± 0.6 mV and represent 100% containment (i.e., confidence) limits for a uniformly distributed error.

I.4.2.3 Measurement Repeatability (ε_{rep})

The LIST-A020 procedure only allows for the collection of two repeat measurements at calibration fixture angle. Therefore, there are insufficient data to evaluate the effects of measurement repeatability.

I.4.2.4   Signal Conditioner Output Equation

The 0 to 4095 counts output range of the SCD-108S corresponds to the positive and negative voltages for angles ranging from - 45° to + 45°. The conversion from volts to counts is equal to 4095 counts/90° × 354°/5V or 3221.4 counts/V. An angle of 0° corresponds to signal conditioner output of 2^{11} or 2048 counts.

The output equation for the signal conditioner module is expressed in equation (I-5). Table I-3 contains relevant information for the equation parameters.

---

[108] SCD-108S Signal Conditioning Card Specifications, www.ttcdas.com.

$$SC_{out} = \left(P_{out} + \varepsilon_{P_{out}} + \varepsilon_{SC_q} \right) \times C_1 + C_2 + \varepsilon_{SC_b} \qquad (I\text{-}5)$$

where

$$P_{out} = V_I \times K \times \frac{X}{X_P} = 5\,\text{V} \times 1 \times \frac{20°}{354°} = 0.282\,\text{V}$$

Table I-3. Parameters used in Signal Conditioner Module Equation

Parameter Name	Description	Nominal or Mean Value	Error Limits	Percent Conf.	Error Distribution
P_{out}	Potentiometer Output	0.282 V			
$\varepsilon_{P_{out}}$	Potentiometer Output Error	0 V			
ε_{SC_b}	Signal Conditioner Bias	0 V	± 20.475 Counts	99	Normal
ε_{SC_q}	Quantization Error	0 V	± 0.6 mV	100	Uniform
C_1	Conversion Coeff.	3,221.4 Counts/V			
$C2$	Conversion Coeff.	2048 Counts			

I.4.3   Data Processor Module (M₃)

The data processing module takes the quantized ADC output and computes a wingboom angle using the linear regression equation (I-1) obtained from calibration data. Errors associated with data processing result from computation round-off or truncation and from residual differences between values observed during calibration and values estimated from the regression equation. Regression error is the primary error source for the data processor module.

I.4.3.1   Regression Error (ε_{reg})

A linear regression equation is typically expressed as

$$\hat{y} = b_0 + bx \qquad (I\text{-}6)$$

where $\hat{y}$ is the predicted value for a given x, b_0 is the value of y when x equals zero, and b represents the amount of change in y with x.

In regression analysis, the standard error of estimate is a measure of the difference between actual values and values estimated from a regression equation.[109] The standard error of estimate is also defined as the standard deviation of the normal distributions of y for any given x.

I.4.3.1.1   Standard Error of Estimate

A regression analysis that has a small standard error of estimate has data points that are very close to the regression line. Conversely, a large standard error of estimate results when data points are widely dispersed around the regression line. The standard error of estimate is computed using equation (I-7).

---

[109] Hanke, J. et al.: *Statistical Decision Models for Management*, Allyn and Bacon, Inc. 1984.

$$s_{y,x} = \sqrt{\frac{\sum(y - \hat{y})^2}{n - 2}} \tag{I-7}$$

The calibration data listed in Table I-1 and the linear regression equation (I-1) were entered into a spreadsheet and the standard error of estimate was computed to be equal to 0.40°.

I.4.3.1.2 Standard Error of Forecast

As previously stated, the standard error of estimate is a measurement of the typical vertical distance of the sample data points from the regression line. However, we must also consider the fact that the regression line was generated from a finite data sample. If another data sample was collected, then a different regression line would result. Therefore, we must also consider the dispersion of various regression lines that would be generated from multiple sample sets around the true population regression line.

The standard error of the forecast accounts for the dispersion of the regression lines and is computed using equation (I-8).

$$s_f = s_{y,x}\sqrt{1 + \frac{1}{n} + \frac{(x - \bar{x})^2}{\sum(x - \bar{x})^2}} \tag{I-8}$$

where $\bar{x}$ is the average or mean of the x values.

The standard error of forecast is computed for each value of x. The wingboom AOA of 20° used in this analysis corresponds to a value of 2930 counts, so s_f has a value of 0.408°. The uncertainty due to regression error is equal to s_f.

I.4.3.2 Data Processor Output Equation

The output equation for the data processing module is expressed in equation (I-9). Table I-4 contains relevant information for the equation parameters.

$$DP_{out} = \left(SC_{out} + \varepsilon_{SC_{out}}\right) \times C_3 + C_4 + \varepsilon_{reg} \tag{I-9}$$

where

$$
\begin{aligned}
SC_{out} &= P_{out} \times C_1 + C_2 \\
&= 0.282\,\text{V} \times 3,221.4\,\text{Counts/V} + 2048\,\text{Counts} \\
&= 908.4\,\text{Counts} + 2048\,\text{Counts} \\
&= 2,956.4\,\text{Counts}
\end{aligned}
$$

Table I-4. Parameters used in Data Processor Module Equation

Parameter Name	Description	Nominal or Mean Value	Standard Uncertainty	Percent Confid.	Error Distribution
SC_{out}	Signal Conditioner Output	2,956.4 Counts			
$\varepsilon_{SC_{out}}$	Signal Conditioner Output Error				

244

C_3	Regression Line Slope	0.0228°/Count			
C_4	Regression Line Intercept	- 45.83°			
ε_{reg}	Regression Error	0°	0.408°		

I.5 Module Error Models

The next step is to develop an error model for each module. Equations (I-3) through (I-5) and (I-9) provide the basis for the development of the module error models.

I.5.1 Potentiometer Module (M_1)

The error model for the potentiometer module is expressed as

$$
\begin{aligned}
\varepsilon_{P_{out}} &= c_{V_I}\varepsilon_{V_I} + c_K\varepsilon_K + c_X\varepsilon_X + c_{X_P}\varepsilon_{X_P} \\
&= c_{V_I}\varepsilon_{V_I} + c_K\left(\varepsilon_L + \varepsilon_R + \varepsilon_N + \varepsilon_{res}\right) + c_{\Delta T}\varepsilon_{TC} + c_X\varepsilon_X + c_{X_P}\varepsilon_{X_P}
\end{aligned}
\tag{I-10}
$$

where c_{V_I}, c_K, c_X, c_{X_P} and $c_{\Delta T}$ are sensitivity coefficients that determine the relative contribution of the error sources to the error in the potentiometer output. The partial derivative equations used to compute the sensitivity coefficients are listed below.

$$
c_{V_I} = \frac{\partial P_{out}}{\partial V_I} = K \times \frac{X}{X_P} \qquad c_K = \frac{\partial P_{out}}{\partial K} = V_I \times \frac{X}{X_P} \qquad c_X = \frac{\partial P_{out}}{\partial X} = \frac{V_I \times K}{X_P}
$$

$$
c_{X_P} = \frac{\partial P_{out}}{\partial X_P} = -\frac{V_I \times K \times X}{X_P^2} \qquad c_{\Delta T} = \frac{\partial P_{out}}{\partial \Delta T} = \frac{\partial P_{out}}{\partial K} \times \frac{\partial K}{\partial \Delta T} = V_I \times \frac{X}{X_P} \times \Delta T
$$

I.5.2 Signal Conditioner Module

The error model for the signal conditioner module is expressed as

$$
\varepsilon_{SC_{out}} = c_{P_{out}}\left(\varepsilon_{P_{out}} + \varepsilon_{SC_q}\right) + c_{SC_b}\varepsilon_{SC_b}
\tag{I-11}
$$

where $c_{P_{out}}$ and c_{SC_b} are sensitivity coefficients that determine the relative contribution of the error sources to the error in the signal conditioner output. The partial derivative equations used to compute the sensitivity coefficients are listed below.

$$
c_{P_{out}} = \frac{\partial SC_{out}}{\partial P_{out}} = C_1 \qquad c_{SC_b} = \frac{\partial SC_{out}}{\partial \varepsilon_{SC_b}} = 1
$$

I.5.3 Data Processor Module

The error model for the data processor module is expressed as

$$
\varepsilon_{DP_{out}} = c_{SC_{out}}\varepsilon_{SC_{out}} + c_{reg}\varepsilon_{reg}
\tag{I-12}
$$

245

where $c_{SC_{out}}$ and c_{reg} are sensitivity coefficients that determine the relative contribution of the error sources to the error in the data processor output. The partial derivative equations used to compute the sensitivity coefficients are listed below.

$$c_{SC_{out}} = \frac{\partial DP_{out}}{\partial SC_{out}} = C_3 \qquad c_{reg} = \frac{\partial SC_{out}}{\partial \varepsilon_{reg}} = 1$$

I.6   Module Uncertainty Models

The next step in the analysis procedure is to develop an uncertainty model for each module, accounting for possible correlations between error sources.

I.6.1   Potentiometer Module

The uncertainty model for the potentiometer module output is developed by applying the variance operator to equation (I-10).

$$u_{\varepsilon_{P_{out}}} = \sqrt{\operatorname{var}\left(\varepsilon_{P_{out}}\right)}$$
$$= \sqrt{\operatorname{var}\left(\begin{array}{l} c_{V_I}\varepsilon_{V_I} + c_K\left(\varepsilon_L + \varepsilon_R + \varepsilon_N + \varepsilon_{res} + \varepsilon_{TC} \times c_{\Delta T}\varepsilon_{\Delta T}\right) \\ +c_X\varepsilon_X + c_{X_P}\varepsilon_{X_P} \end{array}\right)} \qquad \text{(I-13)}$$

There are no correlations between error sources for the potentiometer module. Therefore, the uncertainty in the potentiometer output can be expressed as

$$u_{\varepsilon_{P_{out}}} = \sqrt{\begin{array}{l} c_{V_I}^2 u_{\varepsilon_{V_I}}^2 + c_K^2\left(u_{\varepsilon_L}^2 + u_{\varepsilon_R}^2 + u_{\varepsilon_N}^2 + u_{\varepsilon_{res}}^2\right) + c_K^2 u_{\varepsilon_{TC}}^2 c_{\Delta T}^2 u_{\varepsilon_{\Delta T}}^2 \\ +c_X^2 u_{\varepsilon_X}^2 + c_{X_P}^2 u_{\varepsilon_{X_P}}^2 \end{array}} \qquad \text{(I-14)}$$

I.6.2   Signal Conditioner Module

The uncertainty model for the signal conditioner module output is developed by applying the variance operator to equation (I-11).

$$u_{\varepsilon_{SC_{out}}} = \sqrt{\operatorname{var}\left(\varepsilon_{SC_{out}}\right)}$$
$$= \sqrt{\operatorname{var}\left(c_{P_{out}}\varepsilon_{P_{out}} + c_{P_{out}}\varepsilon_{SC_q} + c_{SC_b}\varepsilon_{SCb}\right)} \qquad \text{(I-15)}$$

There are no correlations between error sources for the signal conditioner module. Therefore, the uncertainty in the signal conditioner output can be expressed as

$$u_{\varepsilon_{SC_{out}}} = \sqrt{c_{P_{out}}^2 u_{\varepsilon_{P_{out}}}^2 + c_{P_{out}}^2 u_{\varepsilon_{SC_q}}^2 + c_{SC_b}^2 u_{\varepsilon_{SC_b}}^2} \qquad \text{(I-16)}$$

I.6.3 Data Processor Module

The uncertainty model for the signal conditioner module output is developed by applying the variance operator to equation (I-12).

$$u_{\varepsilon_{DP_{out}}} = \sqrt{\mathrm{var}\left(\varepsilon_{DP_{out}}\right)}$$

$$= \sqrt{\mathrm{var}\left(c_{SC_{out}}\,\varepsilon_{SC_{out}} + c_{reg}\varepsilon_{reg}\right)} \tag{I-17}$$

There are no correlations between error sources for the data processor module. Therefore, the uncertainty in the data processor output can be expressed as

$$u_{\varepsilon_{DP_{out}}} = \sqrt{c_{SC_{out}}^{2}\,u_{\varepsilon_{SC_{out}}}^{2} + c_{reg}^{2}\,u_{\varepsilon_{reg}}^{2}} \tag{I-18}$$

I.7 Estimate Module Uncertainties

The next step in the system analysis is to estimate uncertainties for the error sources identified for each module and to use these estimates to compute the combined uncertainty and associated degrees of freedom for each module output.

I.7.1 Potentiometer Module

As discussed in section I.4.1, with the exception of resolution error, the error sources identified for the potentiometer module are assumed to follow a normal distribution. Therefore, the corresponding uncertainties can be estimated from the error limits, $\pm L$, confidence level, p, and the inverse normal distribution function, $\Phi^{-1}(\cdot)$, as discussed in Chapter 3.

$$u = \frac{L}{\Phi^{-1}\left(\dfrac{1+p}{2}\right)}$$

For example, the uncertainty due to the supply voltage error is estimated to be

$$u_{\varepsilon_{V_I}} = \frac{0.015\ \mathrm{V}}{\Phi^{-1}\left(\dfrac{1+0.95}{2}\right)} = \frac{0.015\ \mathrm{V}}{1.9600} = 0.00765\ \mathrm{V}.$$

The resolution error follows a uniform distribution, so resolution uncertainty is estimated to be

$$u_{\varepsilon_{res}} = \frac{0.0011}{\sqrt{3}} = \frac{0.0011}{1.732} = 0.000635.$$

The estimated uncertainties for each potentiometer error source are summarized in Table I-5. The component uncertainty for each error source is the positive product of the standard uncertainty and the sensitivity coefficient. The uncertainty in the potentiometer output is computed by taking the root sum square of the component uncertainties.

247

Table I-5. Uncertainty Analysis Results for Potentiometer Module

Error Source	± Error Limits	Error Distribution	Confidence Level (%)	Standard Uncertainty	Sensitivity Coefficient	Component Uncertainty
ε_{V_I}	0.015 V	Normal	95	0.00765 V	0.0565	0.0004 V
ε_X	0.25°	Normal	99	0.0971°	0.01412 V/°	0.0014 V
ε_{X_P}	2°	Normal	95	1.02°	-0.0008 V/°	0.0008 V
ε_L	0.005	Normal	95	0.0026	0.282	0.00072 V
ε_R	0.03	Normal	95	0.0153	0.282	0.0043 V
ε_N	0.02	Normal	95	0.0102	0.282	0.0029 V
ε_{res}	0.0011	Uniform	100	0.000635	0.282	0.00018 V
ε_{TC}	7e^{-5} /C	Normal	95	3.57e^{-5} /°C	14.1 V•°C	0.0005V
Module Output	**0.282 V**				**Output Uncertainty**	**0.0055 V**

The pareto chart, shown in Figure I-4, indicates that uncertainties due to resistance error and noise are the largest contributors to the uncertainty in the potentiometer output.

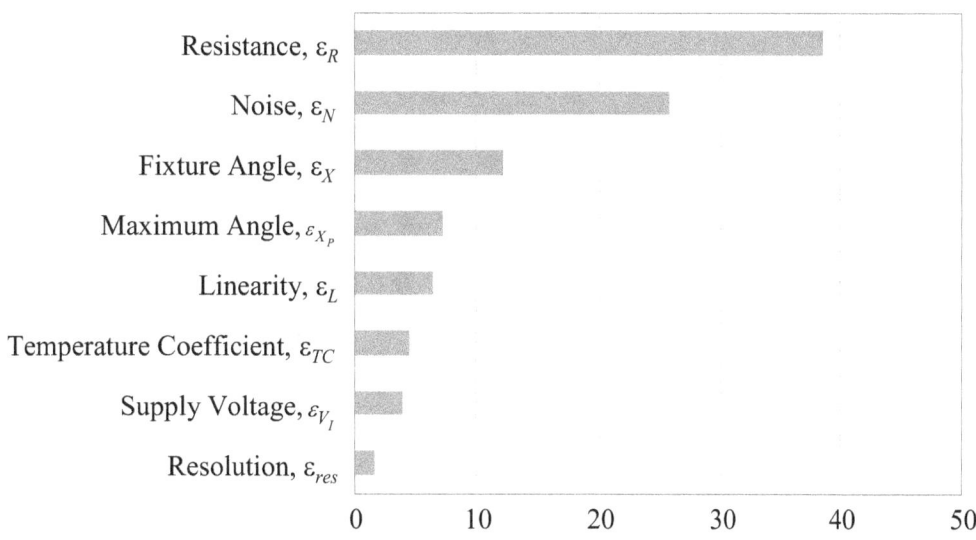

Percent Contribution to Potentiometer Output Uncertainty

Figure I-4. Pareto Chart for Potentiometer Module

I.7.2   Signal Conditioner Module

As discussed in Section I.4.2, the signal conditioner bias is assumed to follow a normal distribution. Therefore, the bias uncertainty is estimated to be

$$u_{\varepsilon_{SC_b}} = \frac{20.475 \text{ Counts}}{\Phi^{-1}\left(\dfrac{1+0.99}{2}\right)} = \frac{20.475 \text{ Counts}}{2.5758} = 7.95 \text{ Counts.}$$

The quantization error follows a uniform distribution, so the associated uncertainty is estimated to be

248

$$u_{\varepsilon_{SC_q}} = \frac{0.0006\,\text{V}}{\sqrt{3}} = \frac{0.0006\,\text{V}}{1.732} = 0.000346\,\text{V}.$$

The estimated uncertainties for the signal conditioner error sources are summarized in Table I-6. The uncertainty in the signal conditioner output is computed by taking the root sum square of the component uncertainties.

Table I-6. Uncertainty Analysis Results for Signal Conditioner Module

Error Source	± Error Limits	Error Distribution	Confidence Level (%)	Standard Uncertainty	Sensitivity Coefficient	Component Uncertainty
$\varepsilon_{P_{out}}$				0.0055 V	3221.4 Counts/V	17.72 Counts
ε_{SC_q}	0.0006 V	Uniform	100	0.000346 V	3221.4 Counts/V	1.12 Counts
ε_{SC_b}	20.475 Counts	Normal	99	7.95 Counts	1	7.95 Counts
Module Output	**2958.0 Counts**				**Output Uncertainty**	**19.45 Counts**

The pareto chart, shown in Figure I-5, indicates that the potentiometer output uncertainty is the largest contributor to the uncertainty in the signal conditioner output.

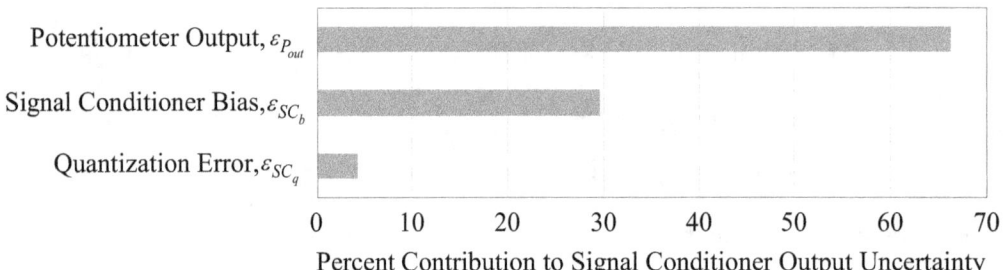

Figure I-5. Pareto Chart for Signal Conditioner Module

I.7.3 Data Processor Module

As discussed in Section I.4.3, the uncertainty due to regression error is equal to the standard error of forecast. For a wingboom angle of 20° used in this analysis, the standard error of forecast was computed to be 0.408°.

The data processor output is

$$
\begin{aligned}
DP_{out} &= SC_{out} \times C_3 + C_4 \\
&= 2{,}956.4\,\text{Counts} \times 0.0228°/\text{Count} - 45.83° \\
&= 67.41° - 45.83° \\
&= 21.6°
\end{aligned}
$$

The estimated uncertainties for the data processor error sources are summarized in Table I-7. The uncertainty in the data processor output is computed by taking the root-sum-square of the component uncertainties.

Table I-7. Uncertainty Analysis Results for Data Processor Module

Error Source	± Error Limits	% Confidence	Standard Uncertainty	Sensitivity Coefficient	Component Uncertainty
$\varepsilon_{SC_{out}}$			19.45 Counts	0.0228°/Count	0.443°
ε_{reg}			0.408°	1	0.408°
Module Output	**21.61°**			**Uncertainty**	**0.603°**

The pareto chart, shown in Figure I-6, indicates that the signal conditioner output uncertainty and regression error uncertainty contribute almost equally to the data processor output uncertainty.

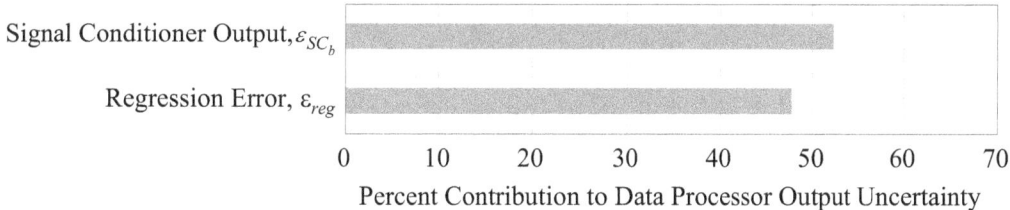

Signal Conditioner Output, ε_{SC_b}

Regression Error, ε_{reg}

Percent Contribution to Data Processor Output Uncertainty

Figure I-6. Pareto Chart for Data Processor Module

I.8   System Output and Uncertainty

In the system analysis approach, each module is analyzed separately and the output and associated uncertainties for each module are propagated to subsequent modules. In the evaluation of the wingboom AOA measurement system modules, it has been illustrated how the uncertainty in the output of one module propagates through to the next module in the series. The module outputs and uncertainties for the wingboom AOA measurement system are summarized in Table I-8.

Table I-8. Summary of Wingboom AOA Measurement System Analysis Results

Module Name	Module Input	Module Output	Standard Uncertainty	Degrees of Freedom
Potentiometer	20°	0.282 V	0.0055 V	∞
Signal Conditioner	0.282 V	2958.0 Counts	19.45 Counts	∞
Data Processor	2958.0 Counts	21.6°	0.60°	∞

The system output and uncertainty are equal to the values computed for the last module in the series. Therefore, the wingboom AOA measurement system has an output of 21.6° with an uncertainty of 0.60°.

REFERENCES

1. Abernathy, R. B. and Ringhauser, B.: "The History and Statistical Development of the New ASME-SAE-AIAA-ISO Measurement Uncertainty Methodology," AIAA/SAE/ASME/ASEE Propulsion Conference, 1985.

2. ANSI/NCSL Z540-2-1997, the *U.S. Guide to the Expression of Uncertainty in Measurement.*

3. Bain, L. J. and Engelhardt, M.: *Introduction to Probability and Mathematical Statistics,* Duxbury Press, 1992.

4. Castrup, H.: "Analytical Metrology SPC Methods for ATE Implementation," presented at the NCSLI Workshop & Symposium, Albuquerque, NM, August 1991.

5. Castrup, H. T. et al.: *NASA Reference Publication 1342, Metrology – Calibration and Measurement Process Guidelines,* Jet Propulsion Laboratory, June 1994.

6. Castrup, H.: "Estimating Category B Degrees of Freedom," presented at the Measurement Science Conference, Anaheim, CA, January 2001.

7. Castrup, H.: "Selecting and Applying Error Distributions in Uncertainty Analysis," presented at the Measurement Science Conference, Anaheim, CA, January 2004.

8. Castrup, H. and Castrup, S.: "Uncertainty Analysis for Alternative Calibration Scenarios," presented at the NCSLI Workshop & Symposium, Orlando, FL, August 2008.

9. Castrup, S.: "A Comprehensive Comparison of Uncertainty Analysis Tools," Proc. Measurement Science Conference, Anaheim, CA, January 2004.

10. Castrup, S.: "Why Spreadsheets are Inadequate for Uncertainty Analysis," Proc. 8th Annual ITEA Instrumentation Workshop, Lancaster, CA, May 2004.

11. Castrup, S.: "Obtaining and Using Equipment Specifications," presented at the NCSLI Workshop & Symposium, Washington, D.C., August 2005.

12. Coleman, H. W. and Steele, W. G.: *Experimentation and Uncertainty Analysis for Engineers,* 2nd Edition, Wiley Interscience Publication, John Wiley & Sons, Inc., 1999.

13. Deaver, David: "Having Confidence in Specifications," proceeding of NCSLI Workshop and Symposium, Salt Lake City, UT, July 2004.

14. Gray, R. M.: *Probability, Random Processes, and Ergodic Properties,* Springer-Verlag 1987. Revised 2001 and 2006-2007 by Robert M. Gray.

15. Hanke, J. et al.: *Statistical Decision Models for Management,* Allyn and Bacon, Inc. 1984.

16. ISA-37.1-1975 - (R1982) *Electrical Transducer Nomenclature and Terminology*, ISA, Research Triangle Park, NC.

17. NASA *Measurement Quality Assurance Handbook* ANNEX 2 – *Measuring and Test Equipment Specifications*.

18. NASA *Measurement Quality Assurance Handbook* ANNEX 4 – *Estimation and Evaluation of Measurement Decision Risk*.

19. National Institute of Science and Technology: *NIST/SEMATECH e-Handbook of Statistical Methods*, http://www.itl.nist.gov/div898/handbook/2005.

20. NCSL, *Establishment and Adjustment of Calibration Intervals*, Recommended Practice RP-1, National Conference of Standards Laboratories, January 1996.

21. Press, et al., *Numerical Recipes in Fortran*, 2nd Ed., Cambridge University Press, 1992.

22. Rice, J.: *Mathematical Statistics and Data Analysis*, Duxbury Press, Belmont, 1995, page 172.

23. Rozanov, Y. A.: *Probability Theory: A Concise Course*, Dover Publications, Inc., 1969.

24. Taylor, J. L.: *Computer-Based Data Acquisition Systems – Design Techniques*, ISA, Research Triangle Park, NC, 1986.

25. Webster, J. G. (Editor-in-Chief): *Measurement, Instrumentation and Sensors Handbook*, Chapman & Hall/CRCnetBase, 1999.

26. Wilson, J. S. (Editor): *Sensor Technology Handbook*, Elsevier Inc., 2005.

www.ingramcontent.com/pod-product-compliance
Lightning Source LLC
Chambersburg PA
CBHW081717220526
45468CB00008B/1872

* 9 7 8 1 7 9 5 5 7 3 3 0 6 *